Lecture Notes in Physics

Springer
Berlin
Heidelberg
New York
Hong Kong
London
Milan
Paris
Tokyo

Physics and Astronomy

ONLINE LIBRARY

http://www.springer.de/phys/

The Editorial Policy for Edited Volumes

The series *Lecture Notes in Physics* (LNP), founded in 1969, reports new developments in physics research and teaching - quickly, informally but with a high degree of quality. Manuscripts to be considered for publication are topical volumes consisting of a limited number of contributions, carefully edited and closely related to each other. Each contribution should contain at least partly original and previously unpublished material, be written in a clear, pedagogical style and aimed at a broader readership, especially graduate students and nonspecialist researchers wishing to familiarize themselves with the topic concerned. For this reason, traditional proceedings cannot be considered for this series though volumes to appear in this series are often based on material presented at conferences, workshops and schools.

Acceptance

A project can only be accepted tentatively for publication, by both the editorial board and the publisher, following thorough examination of the material submitted. The book proposal sent to the publisher should consist at least of a preliminary table of contents outlining the structure of the book together with abstracts of all contributions to be included. Final acceptance is issued by the series editor in charge, in consultation with the publisher, only after receiving the complete manuscript. Final acceptance, possibly requiring minor corrections, usually follows the tentative acceptance unless the final manuscript differs significantly from expectations (project outline). In particular, the series editors are entitled to reject individual contributions if they do not meet the high quality standards of this series. The final manuscript must be ready to print, and should include both an informative introduction and a sufficiently detailed subject index.

Contractual Aspects

Publication in LNP is free of charge. There is no formal contract, no royalties are paid, and no bulk orders are required, although special discounts are offered in this case. The volume editors receive jointly 30 free copies for their personal use and are entitled, as are the contributing authors, to purchase Springer books at a reduced rate. The publisher secures the copyright for each volume. As a rule, no reprints of individual contributions can be supplied.

Manuscript Submission

The manuscript in its final and approved version must be submitted in ready to print form. The corresponding electronic source files are also required for the production process, in particular the online version. Technical assistance in compiling the final manuscript can be provided by the publisher's production editor(s), especially with regard to the publisher's own LaTeX macro package which has been specially designed for this series.

LNP Homepage (http://www.springerlink.com/series/lnp/)

On the LNP homepage you will find:
−The LNP online archive. It contains the full texts (PDF) of all volumes published since 2000. Abstracts, table of contents and prefaces are accessible free of charge to everyone. Information about the availability of printed volumes can be obtained.
−The subscription information. The online archive is free of charge to all subscribers of the printed volumes.
−The editorial contacts, with respect to both scientific and technical matters.
−The author's / editor's instructions.

R. Pastor-Satorras M. Rubi A. Diaz-Guilera (Eds.)

Statistical Mechanics of Complex Networks

Springer

Editors

Romualdo Pastor- Satorras
Universitat Politècnica de Catalunya
Departament de Física
i Enginyeria Nuclear
Campus Nord, Módul B4
08034 Barcelona, Spain

Miguel Rubi
Facultat de Física
Departament de Física Fonamental
Martí i Franquès, 1
08028 Barcelona, Spain

Albert Diaz-Guilera
Facultat de Física
Departament de Física Fonamental
Martí i Franquès, 1
08028 Barcelona, Spain

Cataloging-in-Publication Data applied for

A catalog record for this book is available from the Library of Congress.

Bibliographic information published by Die Deutsche Bibliothek
Die Deutsche Bibliothek lists this publication in the Deutsche Nationalbibliografie;
detailed bibliographic data is available in the Internet at http://dnb.ddb.de

ISSN 0075-8450
ISBN 3-540-40372-8 Springer-Verlag Berlin Heidelberg New York

Springer-Verlag Berlin Heidelberg New York
a member of BertelsmannSpringer Science+Business Media GmbH

http://www.springer.de

© Springer-Verlag Berlin Heidelberg 2003
Printed in Germany

Typesetting: Camera-ready by the authors/editor
Cover design: *design & production*, Heidelberg

Printed on acid-free paper
54/3141/du - 5 4 3 2 1 0

Preface

This volume gathers a number of selected contributions from the XVIII Sitges Conference "Statistical Mechanics of Complex Networks", held on 10–14 June 2002 in Sitges, Barcelona (Spain). The contributions collected provide a general overview of the recent developments in the field of complex networks, concerning in particular their application in physics, biology, and sociology.

The conference was sponsored by institutions that generously provided financial support: DGICYT of the Spanish Government, CIRIT of the Generalitat of Catalonia, Universitat de Barcelona, and Centre Especial de Recerca (CER) Physics of Complex Systems. The city of Sitges allowed us, as usual, to use the Palau Maricel as the lecture hall. We are also very grateful to all those who collaborated in the organization of the event, Drs. C.J. Pérez, A. Pérez-Madrid, M.-C. Miguel, A. Arenas, and Profs. P. Hänggi and A. Vespignani.

Finally, we wish to express out gratitude to all the speakers and participants in the conference, who contributed to create a high scientific level and a very pleasant atmosphere.

Barcelona,   *The Editors*
March 2003

Contents

List of Contributors

Alex Arenas
Departament d'Enginyeria
Informàtica
Universitat Rovira i Virgili
Tarragona, Spain

Albert-László Barabási
Department of Physics
225 Nieuwland Science Hall
University of Notre Dame
Notre Dame, IN 46556, USA

Daniel ben-Avraham
Dept. of Physics
Clarkson University
Potsdam, NY, USA

Marián Boguñá
Department de Física Fonamental
Universitat de Barcelona
Av. Diagonal 647
08028 Barcelona, Spain

Antonio Cabrales
Departament d'Economia i Empresa
Universitat Pompeu Fabra
Barcelona, Spain

Guido Caldarelli
INFM and Dipartimento di Fisica
Università "La Sapienza"
P.le Aldo Moro 5
00185 Roma, Italy

Reuven Cohen
Minerva Center and Dept. of Physics
Bar-Ilan University
Ramat-Gan, Israel

Albert Díaz-Guilera
Departament de Física Fonamental
Universitat de Barcelona
Barcelona, Spain

Ramon Ferrer i Cancho
ICREA-Complex Systems Lab
Universitat Pompeu Fabra (GRIB)
Dr Aiguader 80
08003 Barcelona, Spain

Diego Garlaschelli
INFM and Dipartimento di Fisica
Università "La Sapienza"
P.le Aldo Moro 5
00185 Roma, Italy

M. Girvan
Department of Physics
Cornell University
Ithaca, NY 14853, USA

Igor Goychuk
Universität Augsburg
Institut für Physik
D-86135 Augsburg, Germany

Roger Guimerà
Departament d'Enginyeria Química
Universitat Rovira i Virgili
Tarragona, Spain

Peter Hänggi
Universität Augsburg
Institut für Physik
86135 Augsburg, Germany

Shlomo Havlin
Minerva Center and Dept. of Physics
Bar-Ilan University
Ramat-Gan, Israel

P.L. Krapivsky
Center for BioDynamics, Center for
Polymer Studies and Department
of Physics
Boston University
Boston MA 02215, USA

J.F.F. Mendes
Departamento de Física
Universidade de Aveiro
Campus Universitário de Santiago
3810-193 Aveiro, Portugal

M.E.J. Newman
Department of Physics
University of Michigan
Ann Arbor, MI 48109, USA

Zoltán Oltvai
Department of Pathology
Northwestern University
Chicago, IL 60611, USA

Romualdo Pastor-Satorras
Department de Física i
Enginyeria Nuclear
Universitat Politècnica de Catalunya
Campus Nord
08034 Barcelona, Spain

Luciano Pietronero
INFM and Dipartimento di Fisica
Università "La Sapienza"
P.le Aldo Moro 5
00185 Roma, Italy

Erzsébet Ravasz
Department of Physics
225 Nieuwland Science Hall
University of Notre Dame
Notre Dame, IN 46556, USA

S. Redner
Center for BioDynamics, Center for
Polymer Studies and Department
of Physics
Boston University
Boston MA 02215, USA

Alejandro Fabian Rozenfeld
Minerva Center and Dept. of Physics
Bar-Ilan University
Ramat-Gan, Israel

Gerhard Schmid
Universität Augsburg
Institut für Physik
86135 Augsburg, Germany

Nehemia Schwartz
Minerva Center and Dept. of Physics
Bar-Ilan University
Ramat-Gan, Israel

R.V. Solé
ICREA-Complex Systems Lab
Universitat Pompeu Fabra (GRIB)
Dr Aiguader 80
08003 Barcelona, Spain

H. Eugene Stanley
Center for Polymer Studies and
Department of Physics
Boston University
Boston, MA 02215, USA

Fernando Vega-Redondo
Departament de Fonaments d'Anàlisi
Econòmica
Universitat d'Alacant
Alacant, Spain

Alessandro Vespignani
Laboratoire de Physique Théorique
(UMR du CNRS 8627)
Bâtiment 210, Université de Paris-Sud
91405 ORSAY Cedex, France

Introduction

Statistical physics has faced for long time the challenge to describe and understand large complex systems composed by a heterogeneous set of elements that interact mainly via non-local interactions. In this sense, while the study of their homogeneous counterparts with local interactions has resulted in most significative achievements (consider for example the theory of critical phenomena), heterogeneous systems have usually been the subject of much less attention, due to the intrinsic difficulty that their analytical treatment implies.

Last years have witnessed, however, a renewed interest in the physics of this kind of heterogenous systems, interest born with the realization that they can be mapped into networks, in which the vertices represent the elements and the edges pairwise interactions between elements. This new approach – which finds its roots in the mathematical realm of graph theory – has allowed a first understanding of these systems in terms of *complex networks*, focusing in the study of their topology. While it represents just a first approximation, missing many microscopic properties, this analysis is still able to provide a great deal of information about their topological structure, which has important consequences on the properties of dynamical processes taking place on top of them.

At first instance, the recent availability of powerful computers has lead to a large amount of empirical studies of many real networks. The result of these efforts has been the reconstruction of graph representations of many real technological, social, and biological networks. The statistical analysis of these maps has unveiled the general presence of a very complex and heterogeneous topology, characterized by statistical fluctuations that extend over many orders of magnitude. The main manifestation of this fluctuations is found in the the *degree distribution* (the probability distribution of the number of connections of any vertex), that in most cases exhibits a power-law behavior, lacking any characteristic length scale, and that has led to the definition of the class of *scale-free* networks.

The large scale fluctuations observed in real networks are the typical signature of emergent phenomena, as observed in many complex systems subject to a dynamical evolution. When considering networks from the perspective of complex systems, the attention is placed in the microscopic rules that govern the dynamics of vertices and edges. Since networks are composed by a large number of interacting elements, the detailed characterization of the dynamics of each element is neglected, focusing instead in the understanding of the cooperative phenomena the emerge from their interactions and the statistical laws governing the system. Such approach is analogous to the statistical physics methodology, that has been proved extremely successful in order to link the microscopic dynamics and interactions of matter to the statistical regularities of macroscopic physical systems.

Pursuing this approach, a large amount of research activity has been recently devoted to apply the statistical physics methodology to the study of growing complex networks. In the statistical physics framework, complex networks are viewed as growing systems that evolve in time by adding and removing vertices and edges. This perspective, opposite to the traditional static graph modeling that is at the core of the classical graph theory, allows the identification of some basic models that, while still missing many details, appear to outline the general dynamical theory required to describe the macroscopic properties of natural complex networks. The introduction of the statistical physics approach to the study of complex networks has also provided new techniques and methods to consider the effect of the network topology on different dynamical processes taking place on top of the networks, such as the resilience to damage, and diffusion or searching processes. In this case, well established techniques in statistical physics, such as percolation theory, mean-field methods, or cellular automata simulations, have been fruitfully used to gain a deeper understanding of the general properties of complex networks.

Motivated by the previous considerations, we gathered several leading experts in the field of complex networks for the XVIII Sitges Conference. This book contains a number of selected contributions that will give the reader a general overview of the most recent developments concerning the application of the new theory of complex networks in fields as diverse as physics, biology, and sociology. Among the various aspects covered by the different contributions, we can mention the description of analytical tools to characterize network models, the description of hierarchies and correlations in real complex networks, and the study of dynamical processes such as percolation, searching, or epidemics.

In view of the successes accomplished, and the vast array of new theoretical and practical applications that complex networks offer for the future, we can expect that their study will become a major area of work in the statistical mechanics of the 21st century. We hope that this book will represent a useful introduction to some of the most recent and interesting topics of this emerging field.

1 Rate Equation Approach for Growing Networks

P.L. Krapivsky and S. Redner

Center for BioDynamics, Center for Polymer Studies and Department of Physics, Boston University, Boston MA 02215, USA

Abstract. The rate equations are applied to investigate the structure of growing networks. Within this framework, the degree distribution of a network in which nodes are introduced sequentially and attach to an earlier node of degree k with rate $A_k \sim k^\gamma$ is computed. Very different behaviors arise for $\gamma < 1$, $\gamma = 1$, and $\gamma > 1$. The rate equation approach is extended to determine the joint order-degree distribution, the degree correlations of neighboring nodes, as well as basic global properties. The complete solution for the degree distribution of a finite-size network is outlined. Some unusual properties associated with the most popular node are discussed; these follow simply from the order-degree distribution. Finally, a toy protein interaction network model is investigated, where the network grows by the processes of node duplication and particular form of random mutations. This system exhibits an infinite-order percolation transition, giant sample-specific fluctuations, and a non-universal degree distribution.

1.1 Introduction

In this contribution, we apply tools from statistical physics, in particular, the rate equations, to quantify geometrical properties of evolving networks [1]. The utility of the rate equations have been amply demonstrated for diverse non-equilibrium phenomena, such as aggregation [2], coarsening [3], and epitaxial surface growth [4]. We will argue that the rate equations are a similarly powerful yet simple tool to analyze growing network systems. In addition to providing comprehensive information about the node degree distribution, the rate equations can be readily adapted to treat the joint order-degree distribution, correlations between node degrees, global properties, and a variety of intriguing fluctuation effects.

We will focus on two classes of models. In the first, which we simply term the growing network, nodes are added sequentially and a single link is established between the new node and a pre-existing node according to an attachment rate A_k that depends only on the degree of the "target" node (Fig. 1.1). Here node degree is the number of links that impinge on the node. This appealing model, first introduced by Simon [5] and rediscovered by Barabási and Albert [6], has become extremely fashionable because of its rich phenomenology and timely applications. Examples include the distribution of biological genera, word frequencies, publications, urban populations, income [5, 7], and the link distribution of the world-wide web [8, 9, 10].

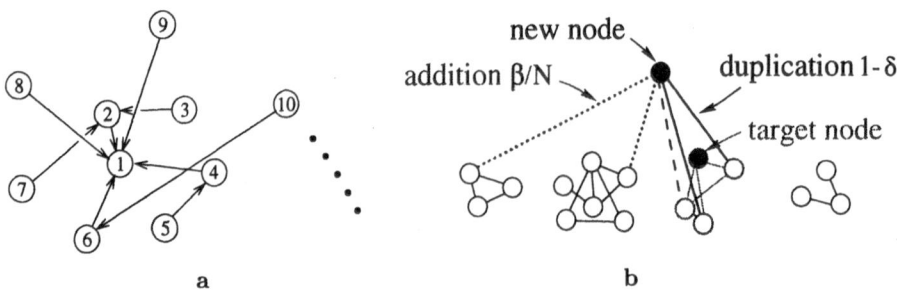

Fig. 1.1. a Growing network. Nodes are added sequentially and a single link joins a new node to an earlier node. Node 1 has degree 5, node 2 has degree 3, nodes 4 and 6 have degree 2, and the remaining nodes have degree 1. **b** Protein interaction network. The new node duplicates 2 out of the 3 links between the target (shaded) and its neighbors. Each successful duplication occurs with probability $1 - \delta$ (*thick solid lines*). The new node also attaches to any other node with probability β/N (*heavy dotted lines*). Thus three previously disconnected clusters are joined by the complete event

The second class of models are inspired by protein interaction networks, where the nodes are individual proteins and the links represent a functional relationship between two proteins in an organism. Much effort has been devoted to infer the structure of such networks [11, 12, 13] and to formulate models that account for their evolution [14, 15, 16, 17, 18, 19]. In the model discussed here [17, 18], nodes are added sequentially and the new node may "duplicate" a randomly chosen target, and the new node can link to any other node with with a small probability (Fig. 1.1). In the duplication step, the new node links to each of the neighbors of the target with probability $1 - \delta$. Thus the duplicate protein is functionally similar to the original [14]. The second process can be viewed as mutation in which a protein can becomes functionally linked to a random subset of other proteins. By this latter process, an arbitrary number of clusters can merge when a single node is introduced. As we shall discuss, this many-body merging leads to an infinite-order percolation transition as a function of the mutation rate. While the applicability of this model to describe real protein networks is still not settled [14], it is a useful starting point for theoretical analysis.

Our basic goal is to quantify the structure of these two basic networks by the rate equation approach.

1.2 Structure of the Growing Network

1.2.1 The Degree Distribution

A fundamental characteristic of any random network is the *node degree distribution* $N_k(N)$, defined as the number of nodes with k links in a network that contains N total nodes. To determine this distribution, we write the rate equations that account for its evolution after each node is introduced. For the growth process in Fig. 1.1a, these rate equations are [20, 21, 22]

$$\frac{dN_k}{dN} = \frac{A_{k-1}N_{k-1} - A_k N_k}{A} + \delta_{k1}. \tag{1.1}$$

The first term on the right, $A_{k-1}N_{k-1}/A$, accounts for processes in which a node with $k-1$ links is connected to the new node, thus increasing N_k by one. Since there are N_{k-1} nodes of degree $k-1$, such processes occur at a rate proportional to $A_{k-1}N_{k-1}$, while the factor $A(N) = \sum_{j\geq1} A_j N_j(N)$ converts this rate into a normalized probability. A corresponding role is played by the second (loss) term on the right-hand side. Here $A_k N_k/A$ is the probability that a node with k links is connected to the new node, thus leading to a loss in N_k. The last term accounts for the introduction of a new node with degree one.

Let us first determine the moments of the degree distribution, $M_n(N) = \sum_{j\geq1} j^n N_j(N)$. Summing (1.1) over all k, gives $\dot{M}_0(N) = 1$. This accords with the definition that $M_0(N) = \sum_k N_k$ is just the total number of nodes N in the network. Similarly, the first moment obeys $\dot{M}_1(N) = 2$, or $M_1(N) = M_1(0)+2N$. Clearly this quantity must grow as $2N$, since introducing a single node creates two link endpoints. Thus the first two moments are *independent* of the attachment kernel A_k and grow linearly in N. On the other hand, higher moments and the degree distribution itself depend in an essential way on A_k.

For general attachment kernels that do not grow faster than linearly with k, it can be easily verified that the asymptotic degree distribution and $A(N)$ both grow linearly with N. Thus substituting $N_k(N) = N\, n_k$ and $A(N) = \mu N$ into (1.1) we obtain the recursion relation $n_k = n_{k-1}A_{k-1}/(\mu + A_k)$ and $n_1 = \mu/(\mu + A_1)$. Solving for n_k, we obtain the formal solution

$$n_k = \frac{\mu}{A_k} \prod_{j=1}^{k} \left(1 + \frac{\mu}{A_j}\right)^{-1}. \tag{1.2}$$

To complete this solution, we need the amplitude μ. Using the definition $\mu = \sum_{j\geq1} A_j n_j$ in (1.2), we obtain the implicit relation

$$\sum_{k=1}^{\infty} \prod_{j=1}^{k} \left(1 + \frac{\mu}{A_j}\right)^{-1} = 1 \tag{1.3}$$

which shows that the amplitude μ depends on the entire attachment kernel.

For the generic case $A_k \sim k^\gamma$, we rewrite the product in (1.2) as the exponential of a sum of logarithms. In the continuum limit, we convert this sum to an integral, expand the logarithm to lowest order, and evaluate the integral to yield:

$$n_k \sim \begin{cases} k^{-\gamma} \exp\left[-\mu\left(\frac{k^{1-\gamma}-2^{1-\gamma}}{1-\gamma}\right)\right], 0 \leq \gamma < 1; \\ k^{-\nu}, \quad \nu = 1+\mu > 2, \qquad\qquad \gamma = 1; \\ \text{singular} \qquad\qquad\qquad\qquad\qquad \gamma > 1. \end{cases} \tag{1.4}$$

That is, for all $0 < \gamma < 1$, the degree distribution is a robust stretched exponential (and pure exponential for $\gamma = 0$). Conversely, for $\gamma > 1$ a phenomenon analogous to gelation occurs in which a single node has almost all of the network links [20, 22]. The regime $\gamma > 1$ actually has an infinite sequence of transitions. For $\gamma > 2$ all but a *finite* number of nodes (in an infinite network) are linked to the "gel" node which has the rest of the links of the network. For $3/2 < \gamma < 2$, the number of nodes with two links grows as $N^{2-\gamma}$, while the number of nodes with more than two links is again finite. For $4/3 < \gamma < 3/2$, the number of nodes with three links grows as $N^{3-2\gamma}$ and the number with more than three is finite. Generally for $(m+1)/m < \gamma < m/(m-1)$, the number of nodes with more than m links is finite, while $N_k \sim N^{k-(k-1)\gamma}$ for $k \leq m$.

The linear kernel ($\gamma = 1$) is on the boundary between these two generic behaviors and leads to a degree distribution that depends on details of the attachment rate. In fact, the exponent $\nu = 1 + \mu$ can be tuned to *any* value larger than 2 [22]. In the special case of the strictly linear kernel, $A_k = k$, the degree distribution has the simple form

$$n_k = \frac{4}{k(k+1)(k+2)} \propto k^{-3}. \tag{1.5}$$

To illustrate the vagaries of asymptotically linear kernels, consider the shifted linear kernel $A_k = k + \lambda$. For this case, note that $A(N) = \sum_j A_j N_j(N)$ gives $A(N) = M_1(N) + \lambda M_0(N)$. Using $A = \mu N$, $M_0 = N$ and $M_1 = 2N$, we get $\mu = 2 + \lambda$. Hence $\nu = 1 + \mu = 3 + \lambda$. Thus an additive shift in the attachment kernel profoundly affects the asymptotic degree distribution. From (1.2), the degree distribution is

$$n_k = (2+\lambda) \frac{\Gamma(3+2\lambda)}{\Gamma(1+\lambda)} \frac{\Gamma(k+\lambda)}{\Gamma(k+3+2\lambda)} \propto k^{-(3+\lambda)}. \tag{1.6}$$

Finally, we discuss a simple extension in which a newly-introduced node links to exactly p earlier nodes [6]. For the linear attachment kernel, the degree distribution $N_k(N)$ (now defined only for $k \geq p$) obeys the rate equation

$$\frac{dN_k}{dN} = \frac{p}{M_1}[(k-1)N_{k-1} - kN_k] + \delta_{k,p}. \tag{1.7}$$

Following the basic approach outlined after (1.3), we find that the asymptotic degree distribution, $n_k = N_k/N$, is [22]

$$n_k = \frac{2p(p+1)}{k(k+1)(k+2)} \qquad \text{for} \quad k \geq p. \tag{1.8}$$

Thus for the strictly linear attachment kernel, the number p of links introduced at each node creation event does not affect the exponent of the degree distribution. Generally, however, this multiple link construction affects the degree distribution. For example, for the shifted linear kernel, we find

$$n_k = \text{const.} \times \frac{\Gamma(k+\lambda)}{\Gamma(k+3+\lambda+\lambda/p)} \quad \text{for } k \geq p,$$

$$n_p = \left(1 + p\,\frac{p+\lambda}{2p+\lambda}\right)^{-1}, \tag{1.9}$$

whose asymptotic behavior is $n_k \sim k^{-(3+\lambda/p)}$. Thus the degree distribution exponent depends strongly on p. This result again shows that fine details of the growth process can be vitally important when the attachment rate is asymptotically linear.

1.2.2 Order Distribution

In addition to node degree, we further characterize a node according to its order of introduction by associating an order index J to the Jth node that was introduced into the network [22, 23]. Let $\mathcal{N}_k(N, J)$ be the probability that the Jth node has degree k when the network has N total nodes. The original degree distribution may be recovered from this joint order-degree distribution through $N_k(N) = \sum_{J=1}^{N} \mathcal{N}_k(N, J)$. The joint distribution evolves according to the rate equation

$$\left(\frac{\partial}{\partial N} - \frac{\partial}{\partial J}\right)\mathcal{N}_k = \frac{A_{k-1}\mathcal{N}_{k-1} - A_k\mathcal{N}_k}{A} + \delta_{k1}\delta(N - J). \tag{1.10}$$

The second term on the left account for the order index evolution. We assume that the probability of linking to a given node depends only on its degree and not on its order.

The homogeneous form of this equation suggests that the solution should depend on the *single* variable $x \equiv J/N$. Writing $\mathcal{N}_k(N, J) = f_k(x)$, converts (1.10) into an ordinary, and readily soluble, differential equation [22]. For the two generic cases of $A_k = k$ and $A_k = 1$, the order-degree distributions are:

$$\mathcal{N}_k(N, J) = \begin{cases} \sqrt{\frac{J}{N}}\left(1 - \sqrt{\frac{J}{N}}\right)^{k-1} & A_k = k, \\[2mm] \dfrac{J}{N}\dfrac{[\ln(N/J)]^{k-1}}{(k-1)!} & A_k = 1. \end{cases} \tag{1.11}$$

For the average order index $\langle J_k \rangle = \sum_k J\,N_k(N, J)/N_k(N)$ of a node of degree k, we find

$$\frac{\langle J_k \rangle}{N} = \begin{cases} \dfrac{12}{(k+3)(k+4)} & A_k = k, \\[2mm] (2/3)^k & A_k = 1. \end{cases} \tag{1.12}$$

Similarly, the average degree $\langle k_J \rangle = \sum_k k\,N_k(N, J)$ of a node of order index J is

$$\langle k_J \rangle = \begin{cases} (J/N)^{-1/2} & A_k = k, \\ \\ \ln(N/J) + 1 & A_k = 1. \end{cases} \tag{1.13}$$

The main messages from these results are that for $A_k = k$, high degree nodes must have been introduced early in the network development. Conversely, for the case of random attachment, $A_k = 1$, high degree nodes could also have been introduced relatively late in the network history. This difference plays a crucial role in determining the properties of the node with the highest degree (Sect. 1.3.2).

1.2.3 Degree Correlations

The rate equation approach also allows us to obtain degree correlations between connected nodes [22]. These develop because a node with large degree is likely to be old [22, 24, 25, 26]. Thus its ancestor is also old and hence has a large degree. To quantify these degree correlations, define $C_{kl}(N)$ as the number of nodes of degree k that attach to an ancestor node of degree l (Fig. 1.2a). For example, in the network of Fig. 1.1, there are $N_1 = 6$ nodes of degree 1, with $C_{12} = C_{13} = C_{15} = 2$. There are also $N_2 = 2$ nodes of degree 2, with $C_{25} = 2$, and $N_3 = 1$ nodes of degree 3, with $C_{35} = 1$.

For simplicity, we consider the linear attachment kernel for which the degree correlation $C_{kl}(N)$ evolves according to

$$M_1 \frac{dC_{kl}}{dN} = (k-1)C_{k-1,l} - kC_{kl} + (l-1)C_{k,l-1} - lC_{kl} + (l-1)C_{l-1} \, \delta_{k1}. \tag{1.14}$$

The processes that gives rise to each term in this equation are illustrated in Fig. 1.3. The first two terms on the right account for the change in C_{kl} due to the addition of a link onto a node of degree $k-1$ (gain) or k (loss) respectively, while the second set of terms gives the change in C_{kl} due to the addition of a link onto the ancestor node. Finally, the last term accounts for the gain in C_{1l} due to the addition of a new node. A crucial feature of this equation is that it is closed; the 2-particle correlation function does not depend on 3-particle quantities.

As in the case of the node degree, the N dependence is simply $C_{kl} = N c_{kl}$. This reduces (1.14) to an N-independent recursion relation. While the details of the solution are unwieldy [22], the asymptotic solution is relatively simple in the scaling regime, $k \to \infty$ and $l \to \infty$ with $y = l/k$ finite:

$$c_{kl} = k^{-4} \frac{4y(y+4)}{(1+y)^4}. \tag{1.15}$$

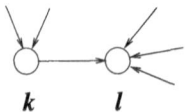

$k \qquad l$

Fig. 1.2. Definition of the node degree correlation C_{kl} for $k = 3$ and $l = 4$

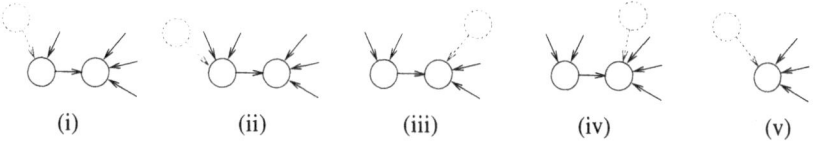

Fig. 1.3. Processes that contribute ((i)–(v) in order) to the terms in the rate equation (1.14) for the case $k = 3$ and $l = 4$ ((i)–(iv)). The newly-introduced node and link are shown dashed. The last case (v) arises only for $k = 1$

For fixed large k, the distribution c_{kl} has a maximum at $y^* = (\sqrt{33} - 5)/2 \cong 0.372$. Thus a node of degree k is typically attached to an ancestor node whose degree is 37% that of the daughter node. In general, when k and l are both large and their ratio is different from one, the limiting behaviors of c_{kl} are

$$c_{kl} \rightarrow \begin{cases} 16\,(l/k^5) & l \ll k, \\ 4/(k^2\,l^2) & l \gg k. \end{cases} \tag{1.16}$$

Here we explicitly see the absence of factorization in the degree correlation: $c_{kl} \neq n_k n_l \propto (k\,l)^{-3}$.

1.2.4 Global Properties

The rate equations can be adapted to determine the *in-component* and *out-component* of the network with respect to a given node **x** [22]. The former is just the set of nodes that point to the node, plus all nodes that refer these daughter nodes, *etc.* The latter are the set of nodes that can be reached by following directed links that emanate from **x** (Fig. 1.4). We study the distribution of these component sizes for the constant attachment kernel, $A_k = 1$, because many results about components are independent of the form of the kernel and thus it suffices to consider the simplest situation.

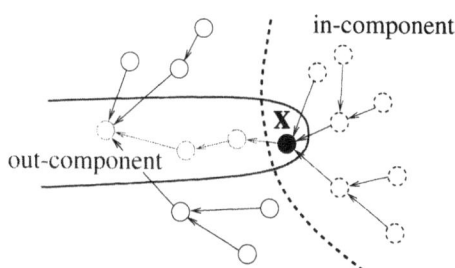

Fig. 1.4. In-component and out-components of node **x**

The In-Component

The number of in-components with s nodes, $I_s(N)$, satisfies the rate equation

$$\frac{dI_s}{dN} = \frac{(s-1)I_{s-1} - sI_s}{A} + \delta_{s1}. \tag{1.17}$$

Here the loss term accounts for processes in which the attachment of a new node to an in-component of size s increases its size by one. This gives a loss rate proportional to s. If there is more than one in-component of size s they must be disjoint, so that the total loss rate for $I_s(N)$ is simply $sI_s(N)$. A similar argument applies for the gain term. Dividing by $A(N) = \sum_j A_j N_j(N)$ converts these rates to probabilities, where $A(N) = M_0(N) \sim N$ for the constant attachment kernel.

It is again easy to verify that each I_s grows linearly in N. Thus we substitute $I_s(N) = N \, i_s$ into (1.17) to obtain $i_s = i_{s-1}(s-1)/(s+1)$ and $i_1 = 1/2$. This immediately gives

$$i_s = \frac{1}{s(s+1)}. \tag{1.18}$$

The s^{-2} tail for the in-component distribution is independent of the form of the attachment kernel [22]. The exponent value also agrees with recent measurements of the web [10].

The Out-Component

The complementary out-component (Fig. 1.4) from each node can be determined by mapping the out-component to an underlying network "genealogy". We build a genealogical tree for the growing network by taking generation $g = 0$ to be the initial node. Nodes that attach to those in generation g are defined to form generation $g + 1$; the node index does not matter in this characterization. For example, in the network of Fig. 1.1a, node 1 is the ancestor of 6, while 10 is the descendant of 6; there are 5 nodes in generation $g = 1$ and 4 in $g = 2$ (Fig. 1.5).

The genealogical tree is convenient because the number O_s of out-components with s nodes equals L_{s-1}, the number of nodes in generation $s - 1$ in the tree

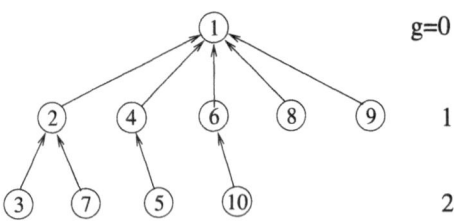

Fig. 1.5. Genealogy of the network in Fig. 1.1a. The nodes indices indicate when each is introduced. The nodes are also arranged according to generation number

(Fig. 1.5). We therefore compute $L_g(N)$, the size of generation g when the network has N total nodes. We again treat the constant attachment kernel; more general cases are treated in [22]. We determine $L_g(N)$ by noting that $L_g(N)$ increases when a new node attaches to a node in generation $g - 1$. This occurs with rate L_{g-1}/M_0, where $M_0(N) = 1 + N$ is the number of nodes. Thus $\dot{L}_g(N) = L_{g-1}/(1+N)$, with solution $L_g(\tau) = \tau^g/g!$, where $\tau = \ln(1+N)$. Thus

$$O_s(\tau) = \tau^{s-1}/(s-1)!. \tag{1.19}$$

The generation size $L_g(N)$ rapidly grows with g for $g < \tau$, and then decreases and becomes of order 1 when $g = e\tau$. To accommodate a network of N nodes, the genealogical tree uses approximately $e\tau$ generations. Therefore the network diameter is $2e\tau \approx 2e\ln N$, since the maximum distance between any pair of nodes is twice the distance from the root to the last generation.

1.3 Finiteness, Fluctuations, and Extremes

1.3.1 Role of Finiteness

Thus far, we have focused on asymptotic properties when the number of nodes is sufficiently large that the ansatz $N_k = N n_k$ is valid. We now consider the role of finiteness on growing networks with attachment rate $A_k = k + \lambda$ ($\lambda > -1$) [27, 28]. This interpolates between linear attachment (for $\lambda = 0$) to random attachment, $A_k = 1$ (for $\lambda \to \infty$).

As quoted in (1.6), the degree distribution of a network with $N \gg 1$ nodes is $N_k(N) \propto N k^{-(3+w)}$ for attachment rate $A_k = k + \lambda$. However, for finite N the degree distribution must eventually deviate from this prediction because the maximal degree cannot exceed N. To establish the range of applicability of (1.6), we estimate the largest degree in the network, $k_{\max}$ by the extreme statistics criterion $\sum_{k \geq k_{\max}} N_k(N) \approx 1$ [29]. This yields $k_{\max} \propto N^{1/(2+\lambda)}$. The degree distribution should therefore deviate from (1.6) when k becomes of the order of $k_{\max}$. The existence of a maximal degree suggests that the degree distribution should have the finite-size scaling form (see also [27, 28, 30, 31, 32])

$$N_k(N) \simeq N n_k F(\xi), \qquad \xi = k/k_{\max}. \tag{1.20}$$

To determine the finite-N behavior of the network, we start by writing the exact recursion relation for the degree distribution after a single node is added:

$$N_k(N+1) = N_k(N) + \frac{(k-1)N_{k-1}(N) - k N_k(N)}{2N}. \tag{1.21}$$

To solve this recursion we introduce the two-variable generating function [28]

$$\mathcal{N}(w, z) = \sum_{N=1}^{\infty} \sum_{k=1}^{\infty} N_k(N) \, w^{N-1} \, z^k,$$

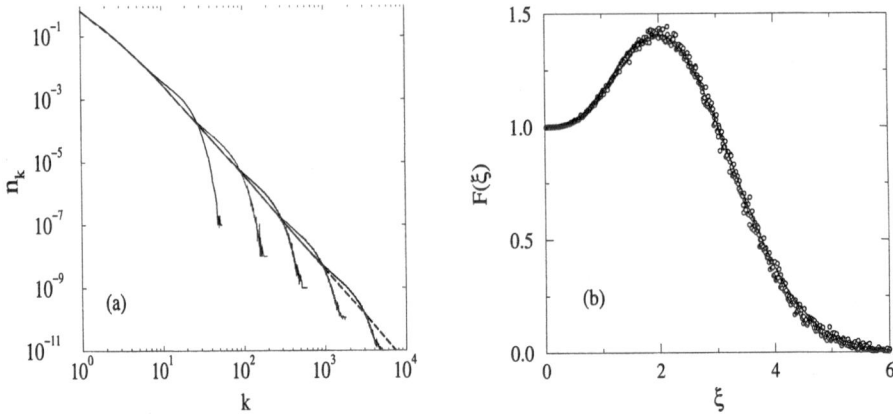

Fig. 1.6. a Normalized degree distribution for networks of $10^2, 10^3, \ldots, 10^6$ nodes (*upper left to lower right*), with 10^5 realizations for each N, for $A_k = k$ for a "triangle" initial condition. The *dashed line* is the asymptotic result $n_k = 4/[k(k+1)(k+2)]$; the last three data sets were averaged over 3, 9, and 27 points, respectively. **b** The corresponding scaling function as defined in $F(\xi)$ in (1.20) from simulation data of 10^6 realizations of a network with $N = 10^4$ nodes for the "dimer" initial condition (*circles*). The *solid curve* (red) is the analytical result of (1.24)

to transform (1.21) into

$$\left(2(1-w)\frac{\partial}{\partial w} + z(1-z)\frac{\partial}{\partial z} - 2\right)\mathcal{N} = \frac{2z}{(1-w)^2}. \qquad (1.22)$$

The exact solution to this equation can be obtained by standard methods and has the unwieldy form [28],

$$\mathcal{N}(w,z) = \frac{(3-2z^{-1})}{(1-w)^2} - \frac{1}{1-w} + \frac{2(z^{-1}-1)}{(1-w)^{3/2}} + \frac{2(1-w)^{-1/2}}{(z^{-1}-1)+(1-w)^{1/2}}$$
$$- \frac{2(z^{-1}-1)^2}{(1-w)^2}\ln\left[1-z+z(1-w)^{1/2}\right]. \qquad (1.23)$$

By expanding $\mathcal{N}(w,z)$, we can determine all the $N_k(N)$. By this approach, we find that the scaling function defined in (1.20) is

$$F(\xi) = \operatorname{erfc}\left(\frac{\xi}{2}\right) + \frac{2\xi+\xi^3}{\sqrt{4\pi}}\,e^{-\xi^2/4}, \qquad (1.24)$$

where $\operatorname{erfc}(x)$ is the complementary error function. A related result was found previously in [27]. This scaling function quantitatively accounts for the large-degree tail of the degree distribution (Fig. 1.6b).

1.3.2 Extremes and Lead Changes

We now investigate properties associated with the statistics of the node with the largest degree – the most popular node [33]; see also [34]. The degree of this node can be determined by a simple extreme statistics argument [29, 33, 34]. Here we discuss related, socially-motivated questions of the identity of the most popular node (the leader). These include the dependence of the leader identity on network size, the rate at which lead changes occur, and the probability that a leader retains the lead as a function of network size.

Leader Identity

We first determine the order index of the leader node. To start with an unambiguous leader, we initialize the system with 3 nodes, with the initial leader having degree 2 (and index 1) and the other two nodes having degree 1. A new leader arises when its degree exceeds that of the current leader. For the linear attachment rate, $A_k = k$, the average order index of the leader $J_{\text{lead}}(N)$ saturates to a finite value of approximately 3.4 as $N \to \infty$ (Fig. 1.7a). With probability ≈ 0.9, the leader is one of the 10 earliest nodes, while the probability that the leader is not among the 30 earliest nodes is less than 0.01. Thus only the earliest nodes have appreciable probabilities to be the leader; the rich really do get richer. In the case of $A_k = k + \lambda$, the average index of the leader also saturates to a finite value that is an increasing function of λ.

For random attachment ($A_k = 1$), the leader index grows as $J_{\text{lead}}(N) \sim N^{\psi}$ with $\psi \approx 0.41$ (Fig. 1.7). The leader is still an early node (since $\psi < 1$), but not necessarily one of the earliest. From our simulations, a node with index greater than 100 has a probability of approximately 10^{-2} of being the leader for

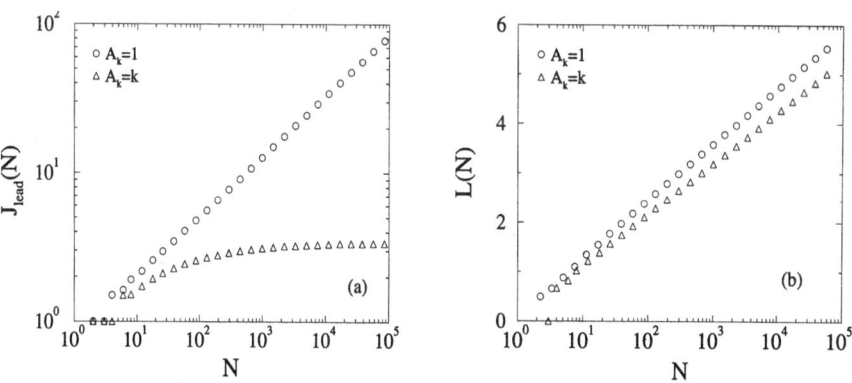

Fig. 1.7. a Average index of the leader $J_{\text{lead}}(N)$ as a function of the total number of nodes N for 10^5 realizations of a growing network. Shown are the cases of attachment rates $A_k = 1$ and $A_k = k$. **b** Average number of lead changes $L(N)$ as a function of network size N for 10^5 realizations of the network for $A_k = 1$ and $A_k = k$

a network of 10^5 nodes. Thus, in random attachment, the order of node creation plays a significant, but not deterministic, role in the identity of the leader node.

For constant attachment rate, the identity of the leader can be simply read off from (1.13); thus the index of the leader node, $J_{\text{lead}}(N) = N(2/3)^{k_{\max}}$ [33]. We estimate the maximum degree from the extremal criterion $\sum_{k \geq k_{\max}} N_k(N) \approx 1$. Using $N_k(N) = N/2^k$, we find $2^{k_{\max}} \approx N$, or $k_{\max} \sim \ln N/\ln 2$. Therefore

$$J_{\text{lead}}(N) \propto N^{\psi}, \quad \text{with} \quad \psi = 2 - \frac{\ln 3}{\ln 2} \approx 0.415\,037,$$

in excellent agreement with our numerical results.

For the linear attachment rate, (1.13) now gives $J_k(N) \sim 12N/k^2$. Since $N_k(N) \sim 4N/k^3$, the extremal criterion $\sum_{k \geq k_{\max}} N_k(N) \approx 1$ now gives $k_{\max} \approx \sqrt{N}$. Therefore $J_{\text{lead}}(N) \sim 12N/k_{\max}^2 = \mathcal{O}(1)$ indeed saturates to a finite value. A similar result holds in the general case $A_k = k + \lambda$. Thus the leader is one of the first few nodes in the network.

Lead Changes

The average number of lead changes $L(N)$ grows logarithmically in N for both $A_k = 1$ and $A_k = k$ (Fig. 1.7), although the details of the underlying distributions of the number of lead changes, $P(L)$, are quite different. For $A_k = 1$, $P(L)$ has a sharp peak, while for $A_k = k$, $P(L)$ has a significant tail that stems from repeated lead changes among the two leading nodes. We also observe numerically that the number of *distinct* nodes that enjoy the lead grows logarithmically in N.

This logarithmic behavior can be easily understood. For $A_k = 1$, the number of lead changes cannot exceed the maximal degree $k_{\max} \sim \ln N/\ln 2$. For the general case $A_k = k + \lambda$, when a new node is added, the lead changes if the leadership is currently shared between two (or more) nodes and the new node attaches to a co-leader. The number of co-leader nodes (with degree $k = k_{\max}$) is $N/k_{\max}^{3+\lambda}$, while the probability of attaching to a co-leader is $k_{\max}/N$. Thus the average number of lead changes satisfies

$$\frac{d}{dN} L(N) \propto \frac{k_{\max}}{N} \frac{N}{k_{\max}^{3+\lambda}}. \tag{1.25}$$

Since $k_{\max}$ grows as $N^{1/(2+\lambda)}$, (1.25) reduces to $dL(N)/dN \propto N^{-1}$ or $L(N) \propto \ln N$. This argument can be adapted to arbitrary attachment rates that do not grow faster than linearly with k.

Fate of the First Leader

Finally, we study the survival probability $S(N)$ that a node that was initially in the lead (has the maximum degree) remains in the lead as the network evolves. For $A_k = k + \lambda$ with $\lambda < \infty$, $S(N)$ is non-zero as $N \to \infty$ (Fig. 1.8). Thus the

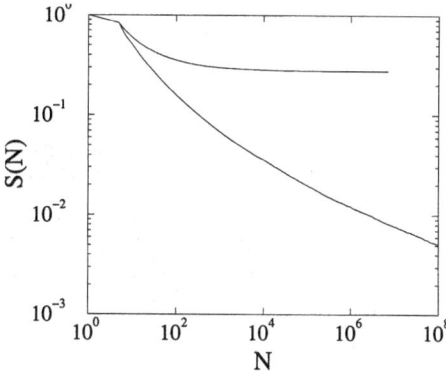

Fig. 1.8. Probability that the first node leads throughout the evolution for 10^5 realizations for $N \leq 10^7$ for $A_k = k$ (*upper*), and $N \leq 10^8$ for $A_k = 1$ (*lower*)

rich get richer holds in a strong form – the lead never changes with a positive probability.

For constant attachment rate the situation is more interesting, as being rich at birth is not as deterministic an influence as in the case of linear attachment. Numerically, $S(N)$ decays very slowly to zero as $N \to \infty$ (Fig. 1.8); a power law $S(N) \propto N^{-\phi}$ is a reasonable fit, but the local exponent is still slowly decreasing at $N \approx 10^8$ where it has reached $\phi(N) \approx 0.18$. To understand this behavior, consider the degree distribution of the first node. This quantity satisfies the recursion relation

$$P(k, N) = \frac{1}{N}P(k-1, N-1) + \frac{N-1}{N}P(k, N-1) \qquad (1.26)$$

which reduces to the convection-diffusion equation

$$\left(\frac{\partial}{\partial \ln N} + \frac{\partial}{\partial k}\right)P = \frac{1}{2}\frac{\partial^2 P}{\partial k^2} \qquad (1.27)$$

in the continuum limit. The solution is a Gaussian

$$P(k, N) = \frac{1}{\sqrt{2\pi \ln N}}\exp\left\{-\frac{(k - \ln N)^2}{2\ln N}\right\}. \qquad (1.28)$$

Thus the degree of the first node grows as $\ln N$, with fluctuations of the order of $\sqrt{\ln N}$. On the other hand, from the degree distribution $N_k(N) = N/2^k$ the maximal degree grows as $k_{\max} = v \ln N$ with $v = 1/\ln 2 \approx 1.44$, and its fluctuations are negligible.

We now estimate $S(N)$ as the probability that the degree of the first node exceeds the maximal degree. For large N, this criterion, $S(N) \approx \sum_{k \geq k_{\max}} P(k, N)$, becomes

$$S(N) \propto \int_{v \ln N}^{\infty} \frac{dk}{\sqrt{\ln N}} \exp\left\{ -\frac{(k - \ln N)^2}{2 \ln N} \right\}$$

$$\propto N^{-\phi} (\ln N)^{-1/2}, \tag{1.29}$$

with $\phi = (v-1)^2/2 \approx 0.0979\ldots$. The recursion (1.26) can, in fact, be solved exactly and gives $P(k, N) = \begin{bmatrix} N \\ k \end{bmatrix}/N!$, for the dimer initial condition, where $\begin{bmatrix} N \\ k \end{bmatrix}$ is the Stirling number of the first kind [35]. Using this instead of the Gaussian approximation leads to the exact exponent $\phi = 1 - v + v \ln v \approx 0.08607$. In either case, the logarithmic factor leads to the very slow approach to asymptotic behavior seen in Fig. 1.8.

1.4 Protein Networks

Finally, we study a toy protein interaction network model that evolves by the biologically-inspired processes of protein duplication and subsequent mutation, as illustrated in Fig. 1.1b [14, 16, 17, 18]. By adapting the rate equation to account for these growth steps, we show that: (i) the system undergoes an infinite-order percolation transition as a function of mutation rate, with a rate-dependent power-law cluster-size distribution everywhere below the threshold, (ii) there are giant fluctuations in network structure and no self-averaging for large duplication rate, and (iii) the degree distribution has a power-law tail with a peculiar rate-dependent exponent.

1.4.1 Infinite-Order Percolation Transition

The protein network has rich percolation properties because the mutation process in Fig. 1.1b can lead to an arbitrary number of clusters being joined in a single step of the evolution. To study these percolation properties, we consider the simpler limit where mutations can occur, but no duplication ($\beta > 0, \delta = 1$). Let $C_s(N)$ be the number of clusters of size $s \geq 1$. This distribution obeys the rate equation

$$\frac{dC_s}{dN} = -\beta \frac{sC_s}{N} + \sum_{n=0}^{\infty} \frac{\beta^n}{n!} e^{-\beta} \sum_{s_1 \cdots s_n} \prod_{j=1}^{n} \frac{s_j C_{s_j}}{N}, \tag{1.30}$$

where the sum is over all $s_1 \geq 1, \ldots, s_n \geq 1$ such that $s_1 + \cdots + s_n + 1 = s$. The first term on the right accounts for the loss of C_s due to the linking of a cluster of size s with the newly-introduced node. The gain term accounts for all possible merging processes of n initially separated clusters whose total size is $s - 1$.

Employing the now familiar ansatz that $C_s(N) = Nc_s$, and introducing the generating function $g(z) = \sum_{s \geq 1} s c_s e^{sz}$, (1.30) becomes

$$g = -\beta g' + (1 + \beta g') e^{z + \beta(g-1)}, \tag{1.31}$$

where $g' = dg/dz$. To detect the percolation transition, we use the fact that $g(0) = \sum s c_s$ is the fraction of nodes within finite clusters. Thus in the non-percolating phase $g(0) = 1$ and the average cluster size $\langle s \rangle = \sum s^2 c_s = g'(0)$, while in the percolating phase the size of the infinite cluster (the giant component) is $NG = N(1 - g(0))$. To determine $g'(0)$, we substitute the expansion $g(z) = 1 + zg'(0) + \ldots$ into (1.31) and take the $z \to 0$ limit. This yields a quadratic equation for $g'(0)$, with solution

$$g'(0) = \langle s \rangle = \frac{1 - 2\beta - \sqrt{1 - 4\beta}}{2\beta^2}. \tag{1.32}$$

This real only for $\beta \leq 1/4$, thus identifying the percolation threshold as $\beta_c = 1/4$. For $\beta > \beta_c$, we express $g'(0)$ in terms of the size of the giant component by setting $z = 0$ in (1.31) to give

$$g'(0) = \frac{e^{-\beta G} + G - 1}{\beta \left(1 - e^{-\beta G}\right)}. \tag{1.33}$$

As $\beta \to \beta_c$, we use $G \to 0$ to simplify (1.33) and find $\langle s \rangle \to (1 - \beta_c)\beta_c^{-2} = 12$. On the other hand, (1.32) shows that $\langle s \rangle \to 4$ when $\beta \to \beta_c$ from below. Thus the average size of finite clusters jumps discontinuously from 4 to 12 as β passes through $\beta_c = 1/4$.

The cluster size distribution c_s exhibits distinct behaviors below, at, and above the percolation transition. For $\beta < \beta_c$, the asymptotic behavior of c_s can be read off from the generating function as $z \to 0$. If c_s has the power-law behavior $c_s \sim B s^{-\tau}$ as $s \to \infty$, then the corresponding generating function $g(z)$ has the small-z expansion $g(z) = 1 + g'(0) z + B\Gamma(2 - \tau)(-z)^{\tau-2} + \ldots$. The regular terms are needed to reproduce the known zeroth and first derivatives of the generating function, while the asymptotic behavior is controlled by the dominant singular term $(-z)^{\tau-2}$. Substituting this expansion into (1.31) we find that the dominant terms are of the order of $(-z)^{\tau-3}$. Balancing all contributions of this order gives

$$\tau = 1 + \frac{2}{1 - \sqrt{1 - 4\beta}}. \tag{1.34}$$

Thus a power-law cluster size distribution with a non-universal exponent arises for *all* $\beta < \beta_c$; that is, the entire range $\beta < \beta_c$ is critical.

At the transition, (1.34) gives $\tau = 3$. However, $c_s \propto s^{-3}$ cannot be correct as it implies that $g'(0)$ diverges. The above expansion of the generating function is also not valid for $\tau = 3$. As in other such situations, we anticipate a logarithmic correction. A detailed analysis of the generating function under this assumption gives [18]

$$c_s \sim \frac{8}{s^3 (\ln s)^2} \quad \text{as} \quad s \to \infty. \tag{1.35}$$

The size of the giant component $G(\beta)$ is obtained by solving (1.31) near $z = 0$. A detailed analysis shows that near β_c

$$G(\beta) \propto \exp\left(-\frac{\pi}{\sqrt{4\beta - 1}}\right), \tag{1.36}$$

so that all derivatives of $G(\beta)$ vanish as $\beta \to \beta_c$. Thus the transition is of infinite order. Similar behaviors were observed [23, 36, 37, 38] for growing network models where single nodes and links were introduced independently. This generic growth mechanism seems to give rise to fundamentally new percolation phenomena.

Giant Fluctuations

In the complementary limit of no mutations ($\beta = 0$), individual realizations of the network evolution fluctuate strongly. We can understand the underlying mechanism for these fluctuations most directly by studying the limit of deterministic duplication ($\delta = 0$), where all the links of the duplicated protein are completed [18]. There is still a stochastic element in this growth, as the node to be duplicated is chosen randomly. Consider the generic initial state of two nodes that are joined by a single link. We denote this graph as $K_{1,1}$, following the graph theoretic terminology [39] that $K_{n,m}$ is the complete bipartite graph in which every node in the subgraph of size n is linked to every node in the subgraph of size m. Duplicating one of the nodes in $K_{1,1}$ gives $K_{2,1}$ or $K_{1,2}$, equiprobably. By continuing to duplicate nodes, it is easy to verify that at every stage the network always remains a complete bipartite graph, say $K_{k,N-k}$, and that every value of $k = 1, \ldots, N - 1$ occurs with equal probability (Fig. 1.9). Thus the degree distribution remains singular – it is always the sum of two delta functions! For fixed N, an average over all realizations of the evolution gives the *average* degree distribution

$$\langle N_k \rangle = 2\left(1 - \frac{k - 1}{N - 1}\right). \tag{1.37}$$

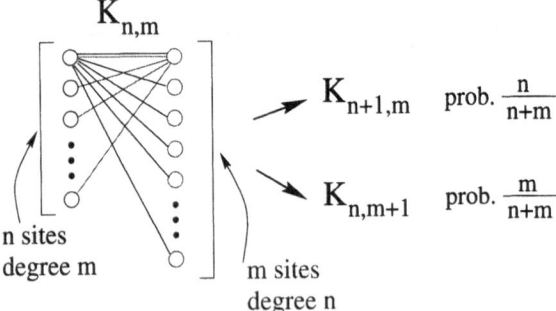

Fig. 1.9. Evolution of the complete bipartite graph $K_{m,n}$ after one deterministic duplication. Only the links emanating from the top nodes of each component are shown

This loss of self averaging is generic; different realizations of the growth lead to statistically distinguishable networks for any initial condition. Similar giant fluctuations also arise in the general case of imperfect duplication where $\delta > 0$ [18].

1.4.2 Non-universal Degree Distribution

Finally, consider the evolution when both incomplete duplication and mutation occur ($\delta < 1$, $\beta > 0$). In each growth step, the average number of links L increases by $\beta + (1 - \delta)\mathcal{D}$ (Fig. 1.1b), where $\mathcal{D}$ is the average node degree of the network. Therefore, $L = [\beta + (1 - \delta)\mathcal{D}]N$. Combining this with $\mathcal{D} = 2L/N$ gives [16, 17]

$$\mathcal{D} = \frac{2\beta}{2\delta - 1}, \tag{1.38}$$

a result that applies only when $\delta > \delta_c = 1/2$. Below this threshold, the number of links grows as

$$\frac{dL}{dN} = \beta + 2(1 - \delta)\frac{L}{N}, \tag{1.39}$$

and combining with $\mathcal{D}(N) = 2L(N)/N$, we find

$$\mathcal{D}(N) = \begin{cases} \text{finite} & \delta > 1/2, \\ \beta \ln N & \delta = 1/2, \\ \text{const.} \times N^{1-2\delta} & \delta < 1/2. \end{cases} \tag{1.40}$$

Without mutation ($\beta = 0$) the average node degree always scales as $N^{1-2\delta}$, so that a realistic finite average degree is recovered *only* when $\delta = 1/2$. Thus mutations play a constructive role, as a finite average degree arises for any duplication rate $\delta > 1/2$.

We now apply the rate equations to study the degree distribution $N_k(N)$ for this case of $\delta > 1/2$ and $\beta > 0$. The degree k of a node increases by one at a rate $A_k = (1 - \delta)k + \beta$. The first term arises because of the contribution from duplication, while mutation leads to the k-independent contribution. The rate equations for the degree distribution are therefore

$$\frac{dN_k}{dN} = \frac{A_{k-1}N_{k-1} - A_k N_k}{N} + G_k. \tag{1.41}$$

The first two terms account for processes in which the node degree increases by one. The source term G_k describes the introduction of a new node of k links, with a of these links created by duplication and $b = k - a$ created by mutation. The probability of the former is $\sum_{s \geq a} n_s \binom{s}{a}(1 - \delta)^a \delta^{s-a}$, where $n_s = N_s/N$ is the probability that a node of degree s is chosen for duplication, while the probability of the latter is $\beta^b e^{-\beta}/b!$. Since duplication and random attachment are independent processes, the source term is

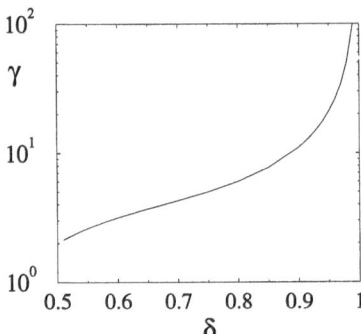

Fig. 1.10. The degree distribution exponent γ as a function of δ from the numerical solution of (1.45)

$$G_k = \sum_{a+b=k} \sum_{s=a}^{\infty} n_s \binom{s}{a} (1-\delta)^a \delta^{s-a} \frac{\beta^b}{b!} e^{-\beta}. \qquad (1.42)$$

Substituting $N_k(N) = N\, n_k$ into the rate equations yields

$$\left(k + \frac{\beta+1}{1-\delta}\right) n_k = \left(k - 1 + \frac{\beta}{1-\delta}\right) n_{k-1} + \frac{G_k}{1-\delta}. \qquad (1.43)$$

Since G_k depends on n_s for all $s \geq k$, the above equation is not a recursion. However, for large k, we reduce it to a recursion by noting that as $k \to \infty$, the main contribution to the sum in (1.42) arises when b is small. Thus a is close to k, and the summand is sharply peaked around $s \approx k/(1-\delta)$. We may then replace the lower limit by $s = k$, and n_s by its value at $s = k/(1-\delta)$. Further, if n_k decays as $k^{-\gamma}$, we write $n_s = (1-\delta)^{\gamma} n_k$ and simplify G_k to

$$G_k \approx (1-\delta)^{\gamma} n_k \sum_{s=k}^{\infty} \binom{s}{k} (1-\delta)^k \delta^{s-k} \sum_{b=0}^{\infty} \frac{\beta^b}{b!} e^{-\beta}$$

$$= (1-\delta)^{\gamma-1} n_k, \qquad (1.44)$$

since the former binomial sum equals $(1-\delta)^{-1}$.

These steps reduce (1.43) to a recursion, from which we deduce that n_k has the power-law behavior $n_k \sim k^{-\gamma}$, with γ determined from [18, 40]

$$\gamma(\delta) = 1 + \frac{1}{1-\delta} - (1-\delta)^{\gamma-2}. \qquad (1.45)$$

The exponent γ has a strong dependence on δ (Fig. 1.10). Further, since the replacement of n_s by $(1-\delta)^{\gamma} n_k$ is valid only asymptotically, the degree distribution should converge slowly to the predicted power law form. This slow approach to asymptotic behavior is observed in large-scale simulations [18]. The corresponding exponent $\gamma(\delta)$ is independent of the mutation rate β but depends sensitively on the duplication rate. Nevertheless, the presence of mutations ($\beta > 0$) is vital to suppress the non-self-averaging as the network evolves and thus make possible a smooth degree distribution.

1.5 Outlook

We hope that the reader is persuaded that the rate equations are a powerful, yet readily applicable tool, to investigate the structure of growing networks. For incrementally growing networks, we have obtained rather complete results for the degree distribution and some of the most important ensuing consequences. We also studied a toy protein interaction network model that evolves by duplication and mutation. In the absence of duplication, the network undergoes an infinite-order percolation transition as a function of the mutation rate. In the absence of mutation, the network exhibits giant sample-specific fluctuations. It is only with the inclusion of mutations that robust and statistically similar networks can be generated.

In summary, the rate equation approach is well-suited to treat a wide range phenomenology associated with evolving networks. Its full potential in this field is just starting to be fully exploited.

The work on protein networks was in collaboration with Byungnam Kahng and Jeenu Kim. This research was supported in part by NSF grant DMR9978902.

References

1. Recent reviews include: S.H. Strogatz: Nature **410**, 268 (2001); R. Albert, A.-L. Barabási: Rev. Mod. Phys. **74**, 47 (2002); S.N. Dorogovtsev, J.F.F. Mendes: Adv. Phys. **51**, 1079 (2002).
2. M.H. Ernst: in *Fractals in Physics*, edited by L. Pietronero, E. Tosatti (Elsevier, Amsterdam, 1986), p. 289.
3. A.J. Bray: Adv. Phys. **43**, 357 (1994).
4. A. Pimpinelli, J. Villain: *Physics of Crystal Growth* (Cambridge University Press, Cambridge, 1998).
5. H.A. Simon: Biometrica **42**, 425 (1955); H.A. Simon: *Models of Man* (Wiley, New York, 1957).
6. A.-L. Barabási, R. Albert: Science **286**, 509 (1999); R. Albert, H. Jeong, A.-L. Barabási: Nature **401**, 130 (1999).
7. G.U. Yule: Phil. Trans. Roy. Soc. B **213**, 21 (1924); *The Statistical Study of Literary Vocabulary* (Cambridge University Press, Cambridge, 1944).
8. S.R. Kumar, P. Raphavan, S. Rajagopalan, A. Tomkins: in *Proc. 8th WWW Conf.* (1999); J. Kleinberg, R. Kumar, P. Raghavan, S. Rajagopalan, A. Tomkins: in Lecture Notes in Computer Science, Vol. 1627 (Springer-Verlag, Berlin, 1999).
9. B.A. Huberman, L.A. Adamic: Nature **401**, 131 (1999); G. Caldarelli, R. Marchetti, L. Pietronero: Europhys. Lett. **52**, 386 (2000)
10. A. Broder, R. Kumar, F. Maghoul, P. Raghavan, S. Rajagopalan, R. Stata, A. Tomkins, J. Wiener: Computer Networks **33**, 309 (2000).
11. P.L. Uetz et al.: Nature **403**, 623 (2000); E.M. Marcotte et al.: Nature **402**, 83 (1999); A. J. Enright et al.: Nature **402**, 86 (1999); T. Ito et al.: Proc. Natl. Acad. Sci. USA **97**, 1143 (2000); ibid **98**, 4569 (2001).
12. J.-C. Rain *et al.*: Nature **409**, 211 (2001).
13. H. Jeong et al.: Nature **411**, 41 (2001).

14. A. Wagner: Mol. Biol. Evol. **18**, 1283 (2001).
15. F. Slanina, M. Kotrla: Phys. Rev. E **62**, 6170 (2000).
16. A. Vazquez, A. Flammini, A. Maritan, A. Vespignani: *cond-mat*/0108043.
17. R.V. Solé, R. Pastor-Satorras, E. D. Smith, T. Kepler: Adv. Complex Syst. **5**, 43 (2002); R. Pastor-Satorras, E.D. Smith, R.V. Solé: preprint.
18. J. Kim, P.L. Krapivsky, B. Kahng, S. Redner: Phys. Rev. E **66**, 055101 (2002).
19. J. Berg, M. Lässig, A. Wagner: *cond-mat*/0207711.
20. P.L. Krapivsky, S. Redner, F. Leyvraz: Phys. Rev. Lett. **85**, 4629 (2000).
21. S.N. Dorogovtsev, J.F.F. Mendes, A.N. Samukhin: Phys. Rev. Lett. **85**, 4633 (2000).
22. P.L. Krapivsky, S. Redner: Phys. Rev. E **63**, 066123 (2001).
23. S.N. Dorogovtsev, J.F.F. Mendes, A. N. Samukhin: Phys. Rev. E **64**, 066110 (2001).
24. M.E.J. Newman: Phys. Rev. Lett. **89**, 208701 (2002).
25. P. Bialas, Z. Burda, J. Jurkiewicz, A. Krzywicki: *cond-mat*/0211527.
26. A. Vazquez: *cond-mat*/0211528.
27. S.N. Dorogovtsev, J.F.F. Mendes, A. N. Samukhin: Phys. Rev. E **63**, 062101 (2001).
28. P.L. Krapivsky, S. Redner: J. Phys. A **35** 9517 (2002).
29. J. Galambos: *The Asymptotic Theory of Extreme Order Statistics* (R.E. Krieger Publishing Co., Malabar, 1987).
30. D.H. Zanette, S.C. Manrubia: Physica A **295**, 1 (2001).
31. L. Kullmann, J. Kertész: Phys. Rev. E **63**, 051112 (2001); D. Lancaster: J. Phys. A **35**, 1179 (2002).
32. Z. Burda, J.D. Correia, A. Krzywicki: Phys. Rev. E **64**, 046118 (2001).
33. P.L. Krapivsky, S. Redner: Phys. Rev. Lett. **xx** xxxx (2002).
34. A.A. Moreira, J.S. de Andrade Jr., L.A.N. Amaral: *cond-mat*/0205411.
35. R.L. Graham, D.E. Knuth, O. Patashnik: *Concrete Mathematics: A Foundation for Computer Science*, (Reading, Mass.: Addison-Wesley, 1989).
36. P.L. Krapivsky, S. Redner: Computer Networks **39**, 261 (2002).
37. D.S. Callaway, J.E. Hopcroft, J.M. Kleinberg, M.E.J. Newman, S.H. Strogatz: Phys. Rev. E **64**, 041902 (2001).
38. M. Bauer, D. Bernard: *cond-mat*/0203232.
39. B. Bollobás: *Modern Graph Theory* (Springer, New York, 1998); S. Janson, T. Luczak, A. Rucinski, *Random Graphs* (Wiley, New York, 2000).
40. F. Chung, L. Lu, T.G. Dewey, D.J. Galas: *cond-mat*/02009008.

2 Directed and Non-directed Scale-Free Networks

Reuven Cohen[1], Alejandro Fabian Rozenfeld[1], Nehemia Schwartz[1],
Daniel ben-Avraham[2], and Shlomo Havlin[1]

[1] Minerva Center and Dept. of Physics, Bar-Ilan University, Ramat-Gan, Israel
[2] Dept. of Physics, Clarkson University, Potsdam, NY, USA

Abstract. Scale-free networks are networks with a scale-free degree distribution, *i.e.*, where the distribution of the number of links per node is a power-law, $p(k) = ck^{-\lambda}$. We review results for the properties of such networks, emphasizing the structural properties of these networks. We begin with normal random scale-free networks and present their percolation properties. We also review results for directed scale-free networks and their percolation properties. Finally we present a study of the possibility of embedding scale-free networks in a lattice.

2.1 Random Scale-Free Networks

The study of random network models began with Erdős and Rényi [1, 2, 3]. They studied models of networks with randomly distributed links. Those models lead to Poisson degree distributions [4]. Due to the development of computers, allowing the analysis of large amounts of data, and the formation of large scale networks, such as the Internet and WWW, some analysis of real world networks has been done in the last decade [5, 6, 7, 8, 9]. This research lead to the conclusion that real world networks are not described correctly by the ER model. The main difference found was that the degree distribution of real world networks studied was found to be very broad rather than the narrow Poisson distribution. Many of the networks studied can be fitted with a scale-free degree distribution. In this chapter we will elaborate on the properties of scale free networks.

A scale free network is a network having a degree distribution:

$$P(k) = ck^{-\lambda}, \tag{2.1}$$

where λ is the exponent and c is an appropriate normalization factor. The distribution is limited by the lower and upper cutoffs, which we will denote by m and K, respectively. The unique properties of this distribution stem from the fact that all moments with $n \geq \lambda - 1$ diverge with K (which is usually increasing with the size of the network).

2.1.1 Percolation Threshold

Percolation theory deals with the cluster structure of networks when a fraction of the sites or bonds is removed. A spanning cluster (or a "giant component" in the terminology of random graphs) is a cluster of connected sites (*i.e.* where there is a path from each site to each other) of the order of the size of the entire network. Most standard treatments of percolation deal with lattices and regular graphs. However, a similar treatment can be applied to random networks.

For a general random network having degree distribution $P(k)$ to have a spanning cluster, a site which is reached by following a link from this cluster must have at least one other link on average to allow the cluster to exist. For this to happen the average degree of a site must be at least 2 (one incoming and one outgoing link) given that the site i is connected to j:

$$\langle k_i|i \leftrightarrow j \rangle = \sum_{k_i} k_i P(k_i|i \leftrightarrow j) = 2. \tag{2.2}$$

Using Bayes rule we get

$$P(k_i|i \leftrightarrow j) = P(k_i, i \leftrightarrow j)/P(i \leftrightarrow j) = P(i \leftrightarrow j|k_i)P(k_i)/P(i \leftrightarrow j), \tag{2.3}$$

where $P(k_i, i \leftrightarrow j)$ is the *joint* probability that node i has degree k_i and that it is connected to node j. For randomly connected networks (neglecting loops) $P(i \leftrightarrow j) = \langle k \rangle/(N-1)$ and $P(i \leftrightarrow j|k_i) = k_i/(N-1)$, where N is the total number of nodes in the network. Using the above criteria (2.2) reduces to [10, 11]:

$$\kappa \equiv \frac{\langle k^2 \rangle}{\langle k \rangle} = 2, \tag{2.4}$$

at the critical point. A spanning cluster exists for graphs with $\kappa > 2$, while graphs with $\kappa < 2$ contain only small clusters whose size is not proportional to that of the entire network. This criterion was derived earlier by Molloy and Reed [10] using a somewhat different arguments.

The negligence of loops can be justified below the threshold since the probability for a bond to form a loop in an s-node cluster is proportional to $(s/N)^2$ (i.e., proportional to the probability of choosing two sites in that cluster). Calculating the fraction of loops P_{loop} in the system yields:

$$P_{loop} \propto \sum_i \frac{s_i^2}{N^2} < \sum_i \frac{s_i S}{N^2} = \frac{S}{N}, \tag{2.5}$$

where the sum is over all clusters in the system and s_i is the size of the ith cluster [12]. Therefore, the fraction of loops in the system is less than or proportional to S/N, where S is the size of the largest cluster. Below the critical threshold there is no spanning cluster in the system and therefore the fraction of loops is negligible. Hence, for values of κ below $\kappa = 2$, loops can be neglected. At the threshold the structure of the spanning cluster is almost a tree. Above the threshold loops can no longer be neglected, but since this only happens when a spanning cluster exists the criterion in (2.4) is valid as a criterion for finding the critical point. A derivation of the exact conditions under which (2.4) is valid can be found in [10].

The above reasoning can be applied to the problem of percolation on a generalized random network. If we randomly remove a fraction p of the sites (or bonds), the degree distribution of the remaining sites will change. For instance, sites with initial degree k_0 will have, after the random removal of nodes, a different number of connections, depending on the number of removed neighbors. The new number of connections will be binomially distributed. If we begin with a distribution of degrees $P_0(k_0)$, the new degree distribution of the network will be:

$$P(k) = \sum_{k_0=k}^{\infty} P_0(k_0) \binom{k_0}{k} (1-p)^k p^{k_0-k}. \tag{2.6}$$

Calculating the first moment for this distribution, given $\langle k_0 \rangle$ and $\langle k_0^2 \rangle$ for the original distribution leads to:

$$\langle k \rangle = \sum_{k=0}^{\infty} P(k)k = (1-p)\langle k_0 \rangle. \tag{2.7}$$

In the same manner we can calculate the second moment:

$$\langle k^2 \rangle = \sum_{k=0}^{\infty} P(k)k^2 = (1-p)^2 \langle k_0^2 \rangle + p(1-p)\langle k_0 \rangle. \tag{2.8}$$

Both those quantities can be substituted into (2.4) to find the criterion for criticality. This yields:

$$\kappa \equiv \frac{\langle k^2 \rangle}{\langle k \rangle} = \frac{(1-p)^2 \langle k_0^2 \rangle + p(1-p)\langle k_0 \rangle}{(1-p)\langle k_0 \rangle} = 2. \tag{2.9}$$

Reorganizing (2.9), one gets the critical threshold for percolation [11]:

$$1 - p_c = \frac{1}{\kappa_0 - 1}, \tag{2.10}$$

where $\kappa_0 \equiv \langle k_0^2 \rangle / \langle k_0 \rangle$ is calculated using the original distribution, before the removal of sites.

Eqations (2.4) and (2.10) are valid for a wide range of generalized random graphs and distributions. For example for a Cayley tree – a graph with a fixed degree z and no loops – the criterion from (2.10) can be used. This yields the critical concentration $q_c = 1 - p_c = 1/(z-1)$, which is well known [13]. Another example is a random Erdös-Rényi (ER) graph. In those graphs edges are distributed randomly and the resulting degree distribution is Poissonian [4]. Applying the criterion from (2.4) to a Poisson distribution yields:

$$\kappa \equiv \frac{\langle k^2 \rangle}{\langle k \rangle} = \frac{\langle k \rangle^2 + \langle k \rangle}{\langle k \rangle} = 2, \tag{2.11}$$

which reduces to $\langle k \rangle = 1$ as known for ER graphs [4].

Our main concern in this chapter will be with the behavior of scale-free networks. Scale-Free networks are networks whose degree distribution (i.e. fraction of sites with k connections) behaves as:

$$P(k) \propto k^{-\lambda}, \ m \leq k \leq K, \tag{2.12}$$

where λ is the exponent, m is the lower cutoff, and K is the upper cutoff. There are no sites with degree below m and above K. For finite networks the upper cutoff N arises naturally since the fraction of high-degree sites decays with k. An estimate of this cutoff can be found by the assumption that the tail of the distribution above K is of the order of one site [11]:

$$\sum_{k=K}^{\infty} P(k) \sim \int_{K}^{\infty} P(k)dk = \frac{1}{N}. \tag{2.13}$$

The estimate obtained this way gives:

$$K \approx mN^{1/(\lambda-1)}. \tag{2.14}$$

This estimate allows the derivation of finite size effects in the network and allows calculations of moments of the distribution in (2.12), that would otherwise diverge. Newman *et al* [14] use an exponential cutoff rather than a sharp one, but the effect on the results is minor.

The importance of scale-free networks lies in the fact that this distribution occurs in many natural and man-made networks [5, 14, 15]. An example of a scale-free network is the physical Internet structure, that is the router to router (and end-units) connectivity. This structure was studied by Faloutsos *et al* [5]. They have found that the inter-router network is a non-directed scale-free network with $\lambda \approx 2.5$. The size of the Internet today is about 10^7 sites, making it a fairly large network.

Further results about the structure of scale-free networks have also been proven by Aiello *et al* [16]. The size of the infinite cluster was calculated, and it was found that for $\lambda \leq 2$ the infinite cluster is of almost the size of the entire graph (i.e. $P_\infty = 1 - o(1)$, where $o(1)$ is a function of the network size, $f(N)$, such that $f(N) \to 0$ when $N \to \infty$). For $\lambda > \lambda_c = 3.478 \ldots$ there is no infinite cluster at all (since we use a somewhat different distribution [17], we get $\lambda_c \approx 4$). For $\lambda < \lambda_c$ the second largest cluster is of order lnN. For lower cutoff $m \geq 2$ a spanning cluster exists for every λ.

The average distance between sites is also different in scale free sites from its value for normal random graphs. While for ER graphs the average distance between sites behaves as $d \sim \ln N$ [4], for scale free graphs with $2 < \lambda < 3$ the distance behaves as $d \sim \ln \ln N$ [18, 19], for $\lambda = 2$, $d \sim const$, and for $\lambda = 3$, $d \sim \ln N/\ln \ln N$ [20]. The reason for this short distance is the small core, containing most high degree sites, which has a very small diameter. For $\lambda > 3$ the random graph behavior $d \sim \ln N$ is recovered. Those results were later confirmed using different methods in [21, 22].

2.1.2 Generating Functions

A general method for studying the size of the infinite cluster and the residual network for a graph with an arbitrary degree distribution was first developed by Molloy and Reed [23]. They suggested viewing the infinite cluster as being explored and used differential equations for the number of un-exposed links and unvisited sites to find the size of the infinite cluster and the degree distribution of the residual graph (the finite clusters).

An alternative and very powerful derivation was given by Newman, Strogatz and Watts [14]. They have used the generating functions method to study the size of the infinite cluster as well as other quantities (such as the diameter and cluster size distribution). They have also applied this method to other types of graphs (directed and bipartite). Here we closely follow their derivation in order to find the size of the infinite cluster and the critical exponents.

In [14, 24] a generating function is built for the degree distribution:

$$G_0(x) = \sum_{k=0}^{\infty} P(k)x^k. \tag{2.15}$$

The probability of reaching a site with degree k by following a specific link is $kP(k)/\langle k \rangle$ [10, 11, 14, 24], and the corresponding generating function for those probabilities is

$$G_1(x) = \frac{\sum kP(k)x^{k-1}}{\sum kP(k)} = \frac{d}{dx}G_0(x)/\langle k \rangle . \tag{2.16}$$

Assuming that $H_1(x)$ is the generating function for the probability of reaching a branch of a given size by following a link, the self-consistent equation for $H_1(x)$ is:

$$H_1(x) = 1 - q + qxG_1(H_1(x)) . \tag{2.17}$$

Since $G_0(x)$ is the generating function for the degree of a site, the generating function for the probability of a site to belong to an n-site cluster is:

$$H_0(x) = 1 - q + qxG_0(H_1(x)) . \tag{2.18}$$

Below the transition, $H_0(1) = 1$, since this is the probability to belong to a cluster of any size. However, above the transition this probability is no longer normalized since this does not include the infinite cluster. Then, the relative size of the giant cluster is $P_\infty = 1 - q + qH_0(1)$, since H_0 contains only the finite-size clusters. It follows that

$$P_\infty = q\left(1 - \sum_{k=0}^{\infty} P(k)u^k\right), \tag{2.19}$$

where $u \equiv H_1(1)$ is the smallest positive root (which can be found numerically) of

$$u = 1 - q + \frac{q}{\langle k \rangle} \sum_{k=0}^{\infty} k P(k) u^{k-1} . \tag{2.20}$$

This equation can be solved numerically and the solution can be substituted into (2.19) to calculate the size of the infinite cluster in a graph with a given degree distribution.

2.1.3 Critical Exponents

Using Abelian and Tauberian methods [25, 26] one can use . (2.19) and (2.20) to find the critical exponents for percolation in scale free networks. Some preliminary results can be found in [27]. A more detailed treatment can be found in [28, 19]. Here we just state the results.

The size of the giant component near the critical point behaves as $P_\infty \sim (p - p_c)^\beta$, where

$$\beta = \begin{cases} \frac{1}{3-\lambda} & 2 < \lambda < 3, \\ \frac{1}{\lambda-3} & 3 < \lambda < 4, \\ 1 & \lambda > 4. \end{cases} \tag{2.21}$$

The number of clusters with size s behaves as $n_s \sim (p - p_c)^{-\tau}$, where

$$\tau = 2 + \frac{1}{\lambda - 2} = \frac{2\lambda - 3}{\lambda - 2} , \qquad 2 < \lambda < 4 . \tag{2.22}$$

For $\lambda > 4$, $\tau = 2.5$, which is the regular mean field value. From those results it can be seen that the critical exponents are anomalous even when the second moment $\langle k^2 \rangle$ is convergent and only the third moment $\langle k^3 \rangle$ diverges, as in the case of $3 < \lambda < 4$.

From τ it can be deduced that the "double jump" in Erdős-Rényi graphs is also seen in scale free graphs, Where S, the size of largest component, scales as $S \sim N^{(\lambda-2)/(\lambda-1)}$ exactly at criticality [19]. For $\lambda \geq 4$ the known result of $S \sim N^{2/3}$ is obtained. The fractal dimensions at criticality for $\lambda > 3$ can also be obtained [19] and are:

$$d_l = \frac{\lambda - 2}{\lambda - 3}, \qquad d_f = 2\frac{\lambda - 2}{\lambda - 3}, \qquad d_c = 2\frac{\lambda - 1}{\lambda - 3}, \tag{2.23}$$

where for $\lambda \geq 4$ the regular mean field values of $2, 4, 6$ are restored.

2.2 Directed Graphs

Many complex networks in nature have directed links, a property that affects the network's navigability and large-scale topology. Here we study the percolation properties of such directed scale-free networks with correlated *in* and *out* degree distributions. We derive a phase diagram that indicates the existence of three regimes, determined by the values of the degree exponents. In the first regime we regain the known directed percolation mean field exponents. In contrast, the second and third regimes are characterized by anomalous exponents, which we calculate analytically. In the third regime the network is resilient to random dilution, i.e., the percolation threshold is $p_c \to 1$.

Recently the topological properties of large complex networks such as the Internet, WWW, electric power grid, cellular and social networks have drawn considerable attention [29, 15]. Some of these networks are directed, for example, in social and economical networks [30] if node A gains information or acquires physical goods from node B, it does not necessarily mean that node B gets similar input from node A. Likewise, most metabolic reactions [31] are one-directional, thus changes in the concentration of molecule A affect the concentration of its product B, but the reverse is not true. Despite the directedness of many real networks, the modeling literature, with few notable exceptions [14, 32], has focused mainly on undirected networks.

An important property of directed networks can be captured by studying their degree distribution, $P(j,k)$, or the probability that an arbitrary node has j incoming and k outgoing edges. Many naturally occurring directed networks, such as the WWW, metabolic networks, citation networks, etc., exhibit a power-law, or *scale-free* degree distribution for the incoming or outgoing links:

$$P_{in(out)}(l) = cl^{-\lambda_{in(out)}}, \quad l \geq m,\qquad(2.24)$$

where m is the minimal connectivity (usually taken to be $m = 1$), c is a normalization factor and $\lambda_{in(out)}$ are the in(out) degree exponents characterizing the network [6, 7]. An important property of scale-free networks is their robustness to random failures, coupled with an increased vulnerability to attacks [33, 11, 24, 27, 34]. Recently it has been recognized that this feature can be addressed analytically in quantitative terms [11, 24, 27] by combining graph theoretical concepts with ideas from percolation theory. Yet, while the percolation properties of undirected networks are much studied, little is known about the effect of node failure in directed networks. As many important networks are directed, it is important to fully understand the implications to their stability. Here we review and extend the results [35] showing that directedness has a strong impact on the percolation properties of complex networks and we draw a detailed phase diagram.

2.2.1 Structure

The structure of a directed graph has been characterized in [14, 32], and in the context of the WWW in [7]. In general, a directed graph consists of a giant weakly connected component (GWCC) and several finite components. In the GWCC every site is reachable from every other, provided that the links are treated as bi-directional. The GWCC is further divided into a giant strongly

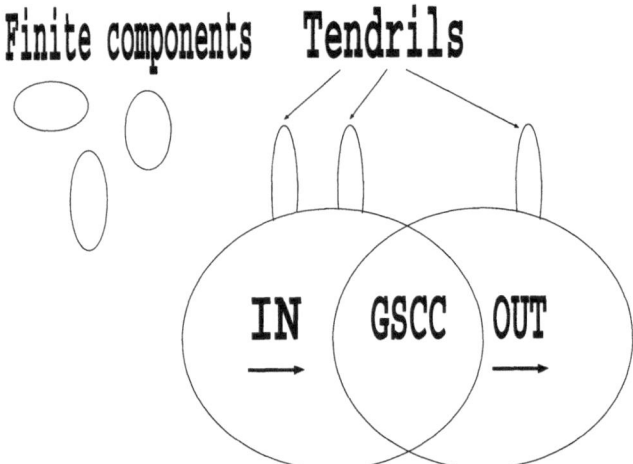

Fig. 2.1. Structure of a general directed graph

connected component (GSCC), consisting of all sites reachable from each other following directed links. All the sites reachable from the GSCC are referred to as the giant OUT component, and the sites from which the GSCC is reachable are referred to as the giant IN component. The GSCC is the intersection of the IN and OUT components. All sites in the GWCC, but not in the IN and OUT components are referred to as the "tendrils" (see Fig. 2.1).

2.2.2 Percolation Threshold

For a directed random network of arbitrary degree distribution the condition for the existence of a giant component can be deduced in a manner similar to [11]. If a site is reached following a link pointing to it, then it must have at least one outgoing link, on average, in order to be part of a giant component. This condition can be written as

$$\langle k_j | i \to j \rangle = \sum_{k_i, k_j} k_j P(k_i, k_j | i \leftrightarrow j) = 1. \tag{2.25}$$

Using Bayes rule we get

$$P(k_i, k_j | i \to j) = \frac{P(k_i, k_j, i \leftrightarrow j)}{P(i \to j)} = \frac{P(i \to j | k_i, k_j) P(k_i, k_j)}{P(i \to j)}. \tag{2.26}$$

For random networks $P(i \to j) = \langle k \rangle / (N-1)$ and $P(i \to j | k_i, k_j) = k_i / (N-1)$, where N is the total number of nodes in the network. The above criterion thus reduces to [14, 32]

$$\langle jk \rangle \geq \langle k \rangle. \tag{2.27}$$

Suppose a fraction p of the nodes is removed from the network. (Alternatively, a fraction $q = 1 - p$ of the nodes is retained.) The original degree distribution, $P(j, k)$, becomes

$$P'(j, k) = \sum_{j_0, k_0}^{\infty} P(j_0, k_0) \binom{j_0}{j} (1 - p)^j p^{j_0 - j} \binom{k_0}{k} (1 - p)^k p^{k_0 - k} . \quad (2.28)$$

In view of this new distribution, (2.27) yields the percolation threshold

$$q_c = 1 - p_c = \frac{\langle k \rangle}{\langle jk \rangle} , \quad (2.29)$$

where averages are computed with respect to the original distribution before dilution, $P(j, k)$. Equation (2.29) indicates that in directed scale-free networks if $\langle jk \rangle$ diverges then $q_c \to 0$ and the network is resilient to random breakdown of nodes and bonds.

The term $\langle jk \rangle$ may be dramatically influenced by the appearance of correlations between the *in-* and *out-*degrees of the nodes. In particular, let us consider scale-free distributions for both the *in-* and *out-*degrees:

$$P_{in}(j) \sim \begin{cases} Bc_{in}j^{-\lambda_{in}} & j \neq 0, \\ 1 - B & j = 0, \end{cases} \quad (2.30)$$

and

$$P_{out}(k) = c_{out}k^{-\lambda_{out}} . \quad (2.31)$$

In (2.30) we choose to add the possible zero value to the *in-*degree in order to maintain $\langle j \rangle = \langle k \rangle$. If the *in-* and *out-*degrees are uncorrelated, we expect $\langle jk \rangle = \langle j \rangle \langle k \rangle$. For several real directed networks this equality does not hold. For example, the network of Notre-Dame University WWW [6], has $\langle k \rangle = \langle j \rangle \approx 4.6$, and thus $\langle j \rangle \langle k \rangle = 21.16$. In contrast, measuring directly we find $\langle jk \rangle \approx 200$, about an order of magnitude larger than the result expected for the uncorrelated case. This yields an estimate of $q_c \approx 0.02$, i.e., a very stable directed network. Similar results are also obtained for metabolic networks studied in [31], indicating that in many real directed networks, the *in-* and *out-*degrees are correlated.

To address correlations, we model it in the following manner: we first generate the j values for the entire network. Next, for each site with $j \neq 0$ with probability A we generate k fully correlated with j, i.e., $k = k(j)$. Assuming that $k(j)$ is a monotonically increasing function then the requirement $c_{out}k^{-\lambda_{out}}dk = c_{in}j^{-\lambda_{in}}dj$ — needed to maintain the distributions scale-free — leads to $k^{\lambda_{out}-1} = j^{\lambda_{in}-1}$. With probability $1 - A$, the degree k is chosen independently from j:

$$P(j, k) \sim \begin{cases} (1 - A)Bc_{in}j^{-\lambda_{in}}c_{out}k^{-\lambda_{out}} + BAc_{out}k^{-\lambda_{out}}\delta_{j, j(k)} & j \neq 0, \\ (1 - B)c_{out}k^{-\lambda_{out}} & j = 0, \end{cases} \quad (2.32)$$

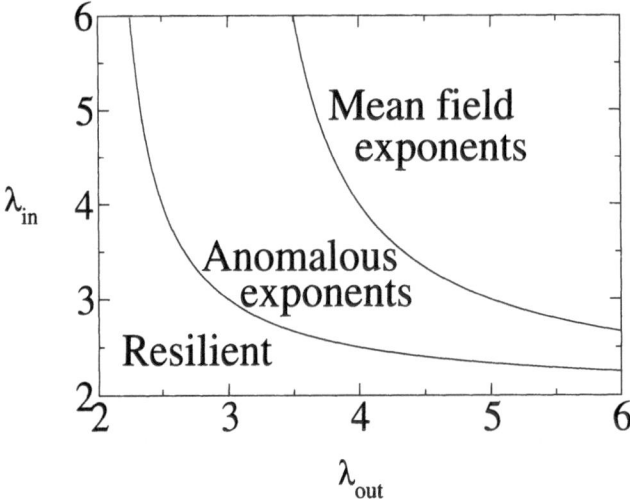

Fig. 2.2. Phase diagram of the different regimes for the IN component of scale-free correlated directed networks. The boundary between Resilient and Anomalous exponents is derived from (2.33) while that between Anomalous exponents and Mean field exponents is given by (2.48) for $\lambda^* = 4$. For the diagram of the OUT component λ_{in} and λ_{out} change roles. After [35]

where $j(k) = k^{\frac{\lambda_{out}-1}{\lambda_{in}-1}}$. With this distribution, any finite fraction BA of fully correlated sites yields a diverging $\langle jk \rangle$ whenever

$$(\lambda_{out} - 2)(\lambda_{in} - 2) \leq 1 , \qquad (2.33)$$

causing the percolation threshold to vanish (see Fig. 2.2). The influence of even very small correlation on the threshold, and the sharpness of the transition to the resilient regime can be seen in Fig. 2.3.

In the case of no correlations between the *in-* and the *out*-degrees, $A = 0$, (2.32) becomes $P(j,k) = P_{in}(j)P_{out}(k)$. Then the condition for the existence of a giant component is: $\langle k \rangle = \langle j \rangle = 1$. Moreover, (2.29) reduces to:

$$q_c = 1 - p_c = \frac{1}{\langle k \rangle} . \qquad (2.34)$$

Applying (2.34) to scale-free networks one concludes that for $\lambda_{out} > 2$ and $\lambda_{in} > 2$ a phase transition exists at a finite q_c. Here we concern ourselves with the critical exponents associated with the percolation transition in both correlated and uncorrelated scale-free network of $\lambda_{out} > 2$ and $\lambda_{in} > 2$, which is the most relevant regime (Fig. 2.2).

Percolation of the GWCC can be seen to be similar to percolation in the non-directed graph created from the directed graph by ignoring the directionality of the links. The threshold is obtained from the criterion [11]

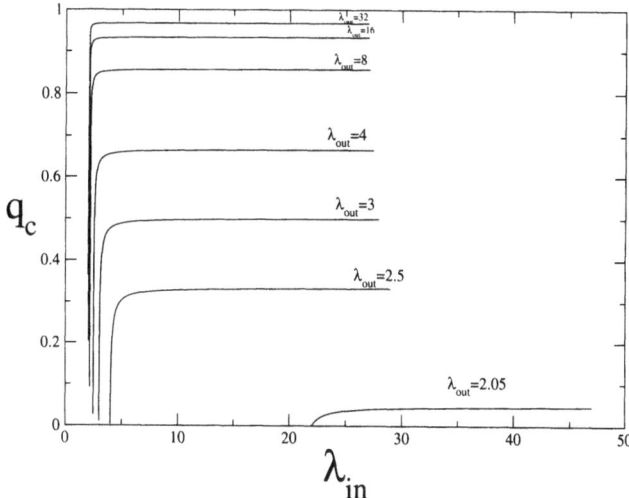

Fig. 2.3. The critical concentration as a function of λ_{in} and λ_{out}. Notice the steep change at the boundaries of the transition between the resilient and non resilient regimes. This plot was obtained for $A = 0.05$

$$q_c = \frac{\langle k \rangle}{\langle k(k-1) \rangle}. \tag{2.35}$$

Here the connectivity distribution is the convolution of the *in* and *out* distributions

$$P'(k) = \sum_{l=0}^{k} P(l, k-l). \tag{2.36}$$

Regardless of correlations, $P'(k)$ is always dominated by the slower decay-exponent, therefore percolation of the GWCC is the same as in non-directed scale-free networks, with $\lambda_{eff} = min(\lambda_{in}, \lambda_{out})$. Note that the percolation threshold of the GWCC may differ from that of the GSCC and the IN and OUT components [32].

2.2.3 Critical Exponents

We now use the formalism of generating functions [26] to analyze percolation of the GSCC and IN and OUT components [35]. In [14, 32] a generating function is built for the joint probability distribution of outgoing and incoming degrees, before dilution:

$$\Phi(x, y) = \sum_{k,j} P(j, k) x^j y^k . \tag{2.37}$$

Using the approach of Callaway *et al* [24], let $q(j, k)$ be the probability that a vertex of degree (j, k) remains in the network following dilution. The generating function after dilution is then

$$G(x, y) = \sum_{k,j} P(j, k)q(j, k)x^j y^k . \tag{2.38}$$

From (2.38) it is possible to define the generating function for the outgoing degrees G_0

$$G_0(y) \equiv G(1, y) = \sum_{k,j} P(j, k)q(j, k)y^k . \tag{2.39}$$

The probability of reaching a site by following a specific link is proportional to $jP(j, k)$, therefore, the probability to reach an occupied site following a specific directed link is generated by

$$G_1(y) = \frac{\sum_{j,k} jP(j, k)q(j, k)y^k}{\sum_{j,k} jP(j, k)} . \tag{2.40}$$

Let $H_1(y)$ be the generating function for the probability of reaching an outgoing component of a given size by following a directed link, after a dilution. $H_1(y)$ satisfies the self-consistent equation:

$$H_1(y) = 1 - G_1(1) + yG_1(H_1(y)) . \tag{2.41}$$

Since $G_0(y)$ is the generating function for the outgoing degree of a site, the generating function for the probability that n sites are reachable from a given site is

$$H_0(y) = 1 - G_0(1) + yG_0(H_1(y)) . \tag{2.42}$$

For the case where correlations exist, and assuming random dilution: $q(j, k) = q$, (2.41) and (2.42) reduce to

$$H_1(y) = 1 - q + \frac{qy}{\langle j \rangle} \sum_k (BAj(k) + (1 - A)\langle j \rangle)P_{out}(k)H_1(y)^k , \tag{2.43}$$

and

$$H_0(y) = 1 - q + qy \sum_k P_{out}(k)H_1(y)^k . \tag{2.44}$$

If $A \to 0$, one expects that $H_0(y) = H_1(y)$, since there is no correlation between j and k, thus the probability to have k outgoing edges is $P_{out}(k)$ whether we choose the site randomly or weighted by the incoming edges j.

$H_0(1)$ is the probability to reach an outgoing component of any *finite* size choosing a site. Thus, below the percolation transition $H_0(1) = 1$, while above the transition there is a finite probability to follow a directed link to a site which is a root of an infinite outgoing component: $P_\infty = 1 - H_0(1)$. It follows that

$$P_\infty(q) = q\left(1 - \sum_k^\infty P_{out}(k)u^k\right), \qquad (2.45)$$

where $u \equiv H_1(1)$ is the smallest positive root of

$$u = 1 - q + \frac{q}{\langle j \rangle} \sum_k (BAj(k) + (1-A)\langle j \rangle)P_{out}(k)u^k. \qquad (2.46)$$

Here $P_\infty(q)$ is the fraction of sites from which an infinite number of sites is reachable. Equation (2.46) can be solved numerically and the solution may be substituted into (2.45), yielding the size of the IN component at dilution $p = 1-q$.

Giant Component Size

Near criticality, the probability to start from a site and reach a giant outgoing component follows $P_\infty \sim (q - q_c)^\beta$. For mean-field systems (such as infinite-dimensional systems, random graphs and Cayley trees) it is known that $\beta = 1$ [36]. This regular mean-field result is not always valid. Instead, we study [35] the behavior of (2.46) near $q = q_c$, $u = 1$, and find

$$\beta = \begin{cases} \frac{1}{3-\lambda^*} & 2 < \lambda^* < 3, \\ \frac{1}{\lambda^*-3} & 3 < \lambda^* < 4, \\ 1 & \lambda^* > 4, \end{cases} \qquad (2.47)$$

where

$$\lambda^* = \lambda_{out} + \frac{\lambda_{in} - \lambda_{out}}{\lambda_{in} - 1}. \qquad (2.48)$$

We see that the order parameter exponent β attains its usual mean-field value only for $\lambda^* > 4$. As $\lambda_{out} \to \lambda_{in}$ the correlated fraction BA of sites resembles non-directed networks [28, 37] (where there is no distinction between incoming and outgoing degrees). In this case we get $\lambda^* = \lambda_{out} = \lambda_{in}$ for any amount of correlation A. The criterion for the existence of a giant component is then $\langle k^2 \rangle / \langle k \rangle = 1$, and not 2 as in the non-directed case. The difference stems from the fact that in the non-directed case one of the links is used to reach the site, while in the directed case there is generally no correlation between the location of the incoming and outgoing links. Therefore, one more outgoing link is available for leaving the site.

Without any correlations, $A = 0$, different terms prevail in the analysis and

$$\beta = \begin{cases} \frac{1}{\lambda_{out}-2} & 2 < \lambda_{out} < 3, \\ 1 & \lambda_{out} > 3. \end{cases} \qquad (2.49)$$

This is the same as (2.47) but with $\lambda^\star = \lambda_{out} + 1$.

The GSCC is the intersection of the IN and OUT components. Therefore, it behaves as the smaller of the two components: $\beta_{GSCC} = max(\beta_{in}, \beta_{out})$. This can be also derived by applying the same methods as for the IN and OUT components to the generating function of the GSCC obtained in [32]. The exponent for the GWCC, on the other hand, is independent of the exponents of the other components, since the transition point is different.

Finite Component Sizes

It is known that for a random graph of arbitrary degree distribution the finite clusters follow the scaling form

$$n(s) \sim s^{-\tau} e^{-s/s^*} , \qquad (2.50)$$

where s is the cluster size and $n(s)$ is the number of clusters of size s. At criticality $s^* \sim |q - q_c|^{-\sigma}$ diverges and the tail of the distribution follows a power law.

The probability that s sites can be reached from a site by following links at criticality follows $p(s) \sim s^{-\tau}$, and is generated by H_0, where $H_0(y) = \sum_s p(s)y^s$. As in [28], $H_0(y)$ can be expanded from (2.42). In the presence of correlations we find [35]

$$\tau = \begin{cases} 1 + \frac{1}{\lambda^\star - 2} & 2 < \lambda^\star < 4, \\ \frac{3}{2} & \lambda^\star > 4. \end{cases} \qquad (2.51)$$

The regular mean-field exponents are recovered for $\lambda^\star > 4$. For the uncorrelated case we get [35]

$$\tau = \begin{cases} 1 + \frac{1}{\lambda_{out} - 1} & 2 < \lambda_{out} < 3, \\ \frac{3}{2} & \lambda_{out} > 3. \end{cases} \qquad (2.52)$$

Now the regular mean-field results are obtained for $\lambda > 3$.

2.2.4 Summary

In summary, we calculate the percolation properties of directed scale-free networks. We find that the percolation critical exponents in scale-free networks are strongly dependent upon the existence of correlations and upon the degree distribution exponents in the range of $2 < \lambda^\star < 4$. This regime characterizes most naturally occurring networks, such as metabolic networks or the WWW. The regular mean-field behavior of percolation in infinite dimensions is recovered only for $\lambda^\star > 4$.

A connection is found between non-directed and directed scale-free percolation exponents for any finite correlation between the *in-* and *out*-degrees. The correlation between the *in-* and *out*-degrees is responsible for the change in the

Table 2.1. Values of λ^* for the different network components for both correlated and uncorrelated cases

	uncorrelated	correlated
GWCC	$min(\lambda_{out}, \lambda_{in}) + 1$	$min(\lambda_{out}, \lambda_{in})$
IN	$\lambda_{out} + 1$	$\lambda_{out} + \frac{\lambda_{in} - \lambda_{out}}{\lambda_{in} - 1}$
OUT	$\lambda_{in} + 1$	$\lambda_{in} + \frac{\lambda_{out} - \lambda_{in}}{\lambda_{out} - 1}$
GSCC	$min(\lambda_{out}, \lambda_{in}) + 1$	$min(\lambda_{out}^*, \lambda_{in}^*)$

critical exponents, and the question whether both incoming and outgoing links lead to the same sites (as in non-directed networks) has no influence on the exponents. In the uncorrelated case, i.e. $P(j, k) = P_{in}(j)P_{out}(k)$, the probability to reach an outgoing component does not bear any dependence upon $P_{in}(j)$. The results are summarized in Table 2.1.

2.3 Spatially Embedded Scale-Free Graphs

The networks studied so far were examples of infinite dimensional networks. They are referred to as infinite dimensional objects since there is no notion of vicinity – every site can connect to every other site with some probability – and since the number of sites in a chemical distance (minimal path length) l from a given site grows exponentially (or faster [18]), which is faster than any power law $N(l) \sim l^d$, expected for a d-dimensional lattice.

Here we describe a method for embedding scale-free networks, with degree distribution $P(k) \sim k^{-\lambda}$, in regular Euclidean lattices accounting for geographical properties [38]. The embedding is driven by a natural constraint of minimization of the total length of the links in the system. All networks with $\lambda > 2$ can be successfully embedded up to an (Euclidean) distance ξ which can be made as large as desired upon the changing of an external parameter. However, the natural cutoff of the distribution can only be achieved for $\lambda > 3$. Clusters of successive layers are found to be compact (the fractal dimension is $d_f = d$), while the dimension of the shortest path between any two sites is smaller than one: $d_{min} = \frac{\lambda - 2}{\lambda - 1 - 1/d}$, contrary to all other known examples of fractals and disordered lattices. An alternative method was suggested by Warren *et al* [39].

All of the networks discussed in previous sections were off-lattice, *i.e.* the Euclidean distance between nodes was irrelevant. However, real-life networks are often embedded in Euclidean geographical space (e.g., the Internet is embedded in the two-dimensional network of routers, neuronal networks are embedded in a three-dimensional brain, etc.). Indeed, in the case of the Internet, indications for the relevance of embedding space is given in [40].

Here we review and extend a method for generating scale-free networks on Euclidean lattices, accounting for geographical properties, and describe some of its properties [38]. As a guiding principle we impose the natural restriction that the total length of links in the system be minimal.

2.3.1 Model Definition

Our model is defined as follows. To each site of a d-dimensional lattice, of size R, and with periodic boundary conditions, we assign a random connectivity k taken from the scale-free distribution

$$P(k) = Ck^{-\lambda}, \qquad m \leq k < K, \tag{2.53}$$

where the normalization constant $C \approx (\lambda - 1)m^{\lambda-1}$ (for K large) [41]. We then select a site at random and connect it to its closest neighbors until its (previously assigned) connectivity k is realized, or until all sites up to a distance

$$r(k) = Ak^{1/d} \tag{2.54}$$

have been explored. (Links to some of the neighboring sites might prove impossible, in case that the connectivity quota of the target site is already filled.) This process is repeated for all sites of the lattice. We show that following this method networks with $\lambda > 2$ can be successfully embedded up to an (Euclidean) distance ξ which can be made as large as desired upon the changing of the external parameter A.

Suppose that one attempts to embed a scale-free network, by the above recipe, in an *infinite* lattice, $R \to \infty$. Sites with a connectivity larger than a certain cutoff $k_c(A)$ cannot be realized, because of saturation of the surrounding sites. Consider the number of links $n(r)$ entering a generic site from a surrounding neighborhood of radius r. Sites at distance r' are linked to the origin with probability $P(k' > (r'/A)^d)$:

$$P\left(k' > \left(\frac{r'}{A}\right)^d\right) = C \int_{(\frac{r'}{A})^d} k^{-\lambda} dk \sim \begin{cases} 1 & r' < A. \\ (\frac{r'}{A})^{d(1-\lambda)} & r' > A. \end{cases} \tag{2.55}$$

Hence

$$n(r) \sim \int_0^r dr' r'^{d-1} P\left(k' > \left(\frac{r'}{A}\right)^d\right) \sim \frac{\lambda-1}{d(\lambda-2)} A^d - \frac{A^{d(\lambda-1)}}{d(\lambda-2)} r^{d(2-\lambda)}. \tag{2.56}$$

The cutoff connectivity is then

$$k_c = \lim_{r \to \infty} n(r) \sim \frac{1}{\lambda-2} A^d. \tag{2.57}$$

The cutoff connectivity implies a cutoff length

$$\xi = r(k_c) \sim (\lambda-2)^{-1/d} A^2. \tag{2.58}$$

The embedded network is *scale-free* up to distances $r < \xi$, and repeats itself (statistically) for $r > \xi$, similar to the infinite percolation cluster above criticality: The infinite cluster in percolation is *fractal* up to the coherence length ξ and repeats thereafter [13, 42, 43].

When the lattice is finite, $R < \infty$, the number of sites is finite, $N \sim R^d$, which imposes a maximum connectivity [11, 44]

$$K \sim mN^{1/(\lambda-1)} \sim R^{d/(\lambda-1)}. \tag{2.59}$$

This implies a finite-size cutoff length

$$r_{max} = r(K) \sim AR^{1/(\lambda-1)}. \tag{2.60}$$

The interplay between the three length scales, R, ξ, r_{max}, determines the nature of the network. If the lattice is finite, then the maximal connectivity is $k_{max} = K$ only if $r_{max} < \xi$. Otherwise ($r_{max} > \xi$) the lattice repeats itself at length scales larger than ξ. As long as $\min(r_{max}, \xi) \ll R$, the finite size of the lattice imposes no serious restrictions. Otherwise ($\min(r_{max}, \xi) \geq R$) finite-size effects become important. We emphasize that in all cases the degree distribution (up to the cutoff) is scale-free.

To study the possibility of embedding the network in the lattice we can use (2.57) in conjunction with (2.54). This yields:

$$r_{max} \equiv r(k_c) = (\lambda - 2)^{1/d} k_c^{2/d} . \tag{2.61}$$

Since we forbid sites to connect further than the lattice size we must demand $r_{max} \leq R = N^{1/d}$, which means that networks can be embedded in a lattice in the suggested manner only if $k_c \leq N^{1/2}$. This limitation imposes an unnatural cutoff whenever $\lambda < 3$, when compared to (2.14).

In Fig. 2.4a we show typical networks that result from our embedding method, for $\lambda = 2.5$ and 5 in two-dimensional lattices (we limit our numerical results to $d = 2$). The larger λ is the more closely the network resembles the embedding lattice, because longer links are rare [45]. In Fig. 2.4b we show the same networks as in part (a) where successive chemical shells are depicted in different colors. Chemical shell l consists of all sites at minimal distance (minimal number of connecting links) l from a given site. For our choice of parameters, $\lambda = 5$ happens to fall in the region of $\xi > r_{max}$, while for $\lambda = 2.5$, $\xi < r_{max}$. In the latter case we clearly see (Fig. 2.4b, $\lambda = 2.5$) the (statistical) repetition of the network beyond the length scale ξ. The different regimes are summarized in Fig. 2.5.

We now address the geometrical properties of the networks, arising from their embedding in Euclidean space. To this aim, it is useful to consider the spatial arrangement of the networks as measured both in an Euclidean metric and in *chemical space*. The chemical distance l between any two sites is the length of the minimal path between them (*minimal* number of links). Thus if the distance between the two sites is r, then $l \sim r^{d_{min}}$ defines the minimal length exponent

$$\lambda = 2.5 \qquad\qquad \lambda = 5$$

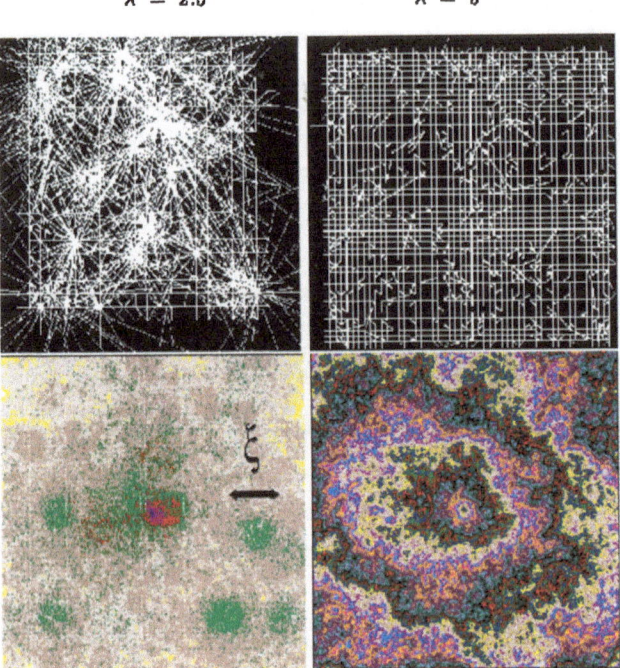

Fig. 2.4. Spatial structure of connectivity network. Top: shown is the typical map of links for a system of 50 x 50 sites generated from a degree distributions with $\lambda = 2.5$ and $\lambda = 5$. Bottom: shown (in different colors) are shells of equidistant sites to the central one in a lattice of 300 x 300 sites. Note that for $\lambda = 5$, shells are concentric and continuous fractals; but for $\lambda = 2.5$, shells are broken

d_{min}. We will see that $d_{min} < 1$ (for $d > 1$), contrary to all naturally occurring fractals and disordered media. Sites at chemical distance l from a given site constitute its l-th chemical shell. The number of (connected) sites within radius r scales as $m(r) \sim r^{d_f}$, defining the fractal dimension d_f. Likewise, the number of (connected) sites within chemical radius l scales as $m(l) \sim l^{d_l}$, which defines the fractal dimension d_l in chemical space. The two fractal dimension are related: $d_{min} = d_f/d_l$ [13, 42, 43].

To study d_f, we compute the perimeter $S(r)$, the number of sites that connect the interior cluster of a region of radius r to sites outside. The fractal dimension then follows from the scaling relation $S(r) \sim r^{d_f-1}$. We focus on the regime $\xi > r_{max}$. Consider a shell dr', of radius r'. A site of connectivity k' within the shell is connected to the outside (to a distance larger than $r - r'$) with probability $P(k' > (\frac{r-r'}{A})^d)$, (2.55). Thus,

$$S(r) = \int_0^r dr'\, r'^{d-1} P\left(k' > \left(\frac{r-r'}{A}\right)^d\right) \sim \begin{cases} r^d & r < A, \\ c(\lambda)Ar^{d-1} & r > A, \end{cases} \quad (2.62)$$

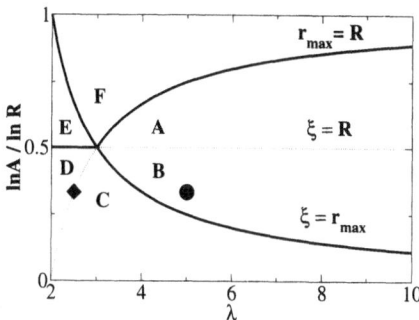

Fig. 2.5. This diagram shows the six regions where different behavior of the network is found: for region A: $r_{max} < R < \xi$, B: $r_{max} < \xi < R$, C: $\xi < r_{max} < R$, D: $\xi < R < r_{max}$, E: $R < \xi < r_{max}$, F: $R < r_{max} < \xi$. The diagram can be mapped into only four regions where the cutoff k_c and where size effect K are expected. A and B: no cutoff and no size effect; C and D: cutoff and no size effect; E: cutoff and size effect; F: no cutoff but size effect. The two symbols indicate the parameters corresponding to Fig. 2.4b, (full diamond) $\lambda = 2.5$ and (full circle) $\lambda = 5$

where $c(\lambda) \sim 1 + 1/[d(\lambda - 1) + 1]$. In other words, the network is compact, $d_f = d$ at large distances $r > A$, and super-compact, $d_f = d + 1$, at $r < A$.

In order to compute d_{min} (or d_l), we regard the chemical shells as being roughly smooth, at least in the regime $\xi > r_{max}$, as suggested by Fig. 2.4b ($\lambda = 5$). Let the width of shell l be $\Delta r(l)$, then

$$l = \int dl = \int \frac{dr}{\Delta r(l)} \sim r^{d_{min}}, \tag{2.63}$$

since $\Delta l = 1$. The number of sites in shell l, $N(l)$, is, on the one hand, $N(l) \sim r(l)^{d-1} \Delta r(l)$. On the other hand, since the maximal connectivity in shell l is $K(l) \sim N(l)^{1/(\lambda-1)}$, the thickness of shell $(l+1)$ is $\Delta r(l+1)$ which is determined by the length of the largest link to the next shell i.e., $r[K(l)]$, and thus, $\Delta r(l+1) \sim r[K(l)] \sim AK(l)^{1/d}$. Assuming (for large l) that $\Delta r(l+1) \sim \Delta r(l)$, we obtain

$$\Delta r(l) \sim r^{\frac{d-1}{d(\lambda-1)-1}}. \tag{2.64}$$

Using this expression in (2.63), yields

$$d_{min} = \frac{\lambda - 2}{\lambda - 1 - 1/d}. \tag{2.65}$$

Thus, above $d = 1$, the dimensions d_{min} and $d_l = d_f/d_{min}$ are anomalous for all values of λ.

In Fig. 2.6 we plot d_{min} as measured from simulations, and compared with the analytical result (2.65). The scaling suggested in Fig. 2.6b, $N(l) \sim$

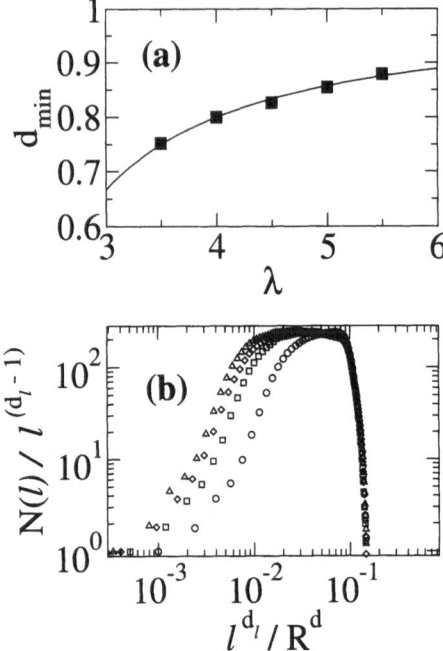

Fig. 2.6. a The minimal length exponent d_{min} as a function of λ. Note the good agreement between theoretical estimations (continuous line) and simulations results (full squares). **b** The shape of the $\Phi(l^{d_l}/R^d)$ scaling function is shown for $\lambda = 4$ and several lattice sizes: R=1000 (circle), 2000 (square), 2500 (diamond) and 3000 (triangle)

$l^{d_l-1}\Phi(l^{d_l}/R^d)$, is valid only for $\xi > r_{max}$. For $R \to \infty$, we expect that the network is scale-free up to length scale ξ and the analogous scaling will be $N(l) \sim l^{d_l-1}\Psi(l^{d_l}/\xi^d)$, where $\Psi(x \gg 1) \sim x^{(d-d_l)/d_l}$.

Note on the Upper Cutoff

In (2.14) we suggest that the upper cutoff of a scale free network scales as $N^{1/(\lambda-1)}$. However, for the spatially embedded graphs we find that no graph with $\lambda < 3$ can be embedded in a lattice without sacrificing the natural cutoff (see discussion after (2.61)). That is, the cutoff is limited to $\sqrt{N}$. This holds true for every d. Similar results are indeed obtained for mean field (*i.e.* non-embedded) graphs [46], while Warren *et al* [39] find the natural cutoff even for graphs embedded in $d = 2$ lattices.

A possible explanation is in the different method for the network implementation, which leads to different ensembles. For the non-embedded networks we allowed every link to lead to every other with an equal probability, thus allowing more than one edge between a pair of sites, and edges leading from a site to itself which were just ignored. In contrast, in the spatially embedded case no such connections were allowed. It is plausible that allowing such connections

$\lambda = 2.0$ $\qquad\qquad$ $\lambda = 2.5$

$\lambda = 5.0$ $\qquad\qquad$ $\lambda = 50$

Fig. 2.7. The infinite cluster in scale free networks at criticality. The clusters were generated using a Leath type method, where the nearest available nodes are selected in each shell

or, alternatively, allowing a deviation from the degree distribution, leads to the "natural" cutoff, while requiring the exact degree sequence in conjunction with no such connections influences the ensemble, bringing to an upper cutoff of $\sqrt{N}$, due to the high probability of forming such connections when the cutoff is higher. The limit of $K \sim \sqrt{N}$ seems to stem from the fact that the expected number of edges between two such sites (or self-loops of a single such site) is of order $K^2/\langle k \rangle N \gg 1$, which implies that most networks having such high degree sites will be multigraphs, and therefore this might limit the cutoff. On the other hand, since degree 1 sites consist of a finite fraction of the links in the network, a finite fraction of the links of high degree sites will link to them, implying that the tail of the distribution, and therefore the scaling of the cutoff is not changed, even when double edges and self loops are removed.

2.3.2 Summary

In summary, we propose a method for embedding scale-free networks in Euclidean lattices. The method is based on a natural principle of minimizing the total length of links in the system. This principle enables us to embed the scale-free networks in Euclidean space without additional external exponents. Very recently, independently, Manna and Sen [47] and Xulvi-Brunet and Sokolov [48] suggested a different embedding method in Euclidean space which include an external exponent. We have shown that while the fractal dimension d_f of the network is the same as the Euclidean dimension, the chemical dimension $d_l > d_f$ for all values of λ, yielding $d_{min} < 1$ for all λ and $d > 1$. A related work by Warren, Sander and Sokolov [39], studies some percolation properties of a similar geographical model. In Fig. 2.7 we show some snapshots of the infinite cluster at the percolation threshold, for $2d$ scale free systems with various values of λ.

Acknowledgments

Daniel ben-Avraham wishes to thank the NSF for support under contract PHY-0140094.

References

1. P. Erdős, A. Rényi: Publicationes Mathematicae **6**, 290 (1959)
2. P. Erdős, A. and Rényi, Publications of the Mathematical Institute of the Hungarian Academy of Sciences **5**, 17 (1960)
3. P. Erdős, A. Rényi: Acta Mathematica Scientia Hungary **12**, 261 (1961)
4. B. Bollobás, *Random Graphs* (Academic Press, London, 1985)
5. M. Faloutsos, P. Faloutsos, C. Faloutsos: ACM SIGCOMM '99 Comput. Commun. Rev. 29, 251 (1999)
6. A.-L. Barabási, R. Albert, H. Jeong: Physica A, **281**, 2115 (2000)
7. A. Broder, R. Kumar, F. Maghoul, P. Raghavan, S. Rajagopalan, R. Stata, A. Tomkins, J. Wiener: Computer Networks **33**, 309 (2000)
8. A.L. Barabási, R. Albert: Science, **286**, 509 (1999)
9. A. Vazquez, R. Pastor-Satorras, A. Vespignani: cond-mat/0206084
10. M. Molloy, B. Reed: Random Structures and Algorithms **6**, 161 (1995)
11. R. Cohen, K. Erez, D. ben-Avraham, S. Havlin: Phys. Rev. Lett. **85**, 4626 (2000)
12. The consideration ignored the degree distribution. However similar considerations should apply to all well behaved distributions
13. A. Bunde, and S. Havlin (editors), *Fractals and Disordered System* (Springer, New York, 1996)
14. M.E.J. Newman, S.H. Strogatz, D.J. Watts: Phys. Rev. E, **64**, 026118 (2001)
15. S.N. Dorogovtsev, J.F.F. Mendes: Adv. in Phys., 51 (4), (2002)
16. W. Aiello, F. Chung, L. Lu: Proc. 32nd ACM Symp. Theor. Comp., (2000)
17. R. Cohen, K. Erez, D. ben-Avraham, S. Havlin: Phys. Rev. Lett. **87**, 219802 (2001)
18. R. Cohen, S. Havlin: cond-mat/0205476
19. R. Cohen, S. Havlin, D. ben-Avraham: "Structural properties of scale free networks", in *Handbook of graphs and networks*, eds. S. Bornholdt and H.G. Schuster (Wiley-VCH, Berlin, 2002)
20. B. Bollobas, O. Riordan: in *Handbook of graphs and networks*, eds. S. Bornholdt and H.G. Schuster (Wiley-VCH, Berlin, 2002)
21. S.N. Dorogovtsev, J.F.F. Mendes, A.N. Samukhin: cond-mat/0210085.
22. F.R.K. Chung, L. Lu: PNAS **99**, 15879 (2002)

23. M. Molloy, B. Reed: Combinatorics, Probability and Computing **7**, 295 (1998)
24. D.S. Callaway, M.E.J. Newman, S.H. Strogatz, D.J. Watts: Phys. Rev. Lett. **85**, 5468 (2000)
25. G.H. Weiss: *Aspects and Applications of the Random Walk* (North-Holland, Amsterdam, 1994)
26. H.S. Wilf: *Generatingfunctionology* 2nd ed. (Academic Press, London, 1994)
27. R. Cohen, K. Erez, D. ben-Avraham, S. Havlin: Phys. Rev. Lett. **86**, 3682 (2001)
28. R. Cohen, D. ben-Avraham, S. Havlin: Phys. Rev. E **66**, 036113 (2002)
29. R. Albert, A.-L. Barabasi: Rev. of Mod. Phys. **74**, 47 (2002)
30. A. Aleksiejuk, J.A. Holyst, G. Kossinets: Int. J. Mod. Phys. C **13**, 333 (2002)
31. H. Jeong, B. Tombor, R. Albert, Z.N. Oltvai, A.-L. and Barabási: Nature, **407**, 651 (2000)
32. S.N. Dorogovtsev, J.F.F. Mendes, A.N. Samukhin: Phys. Rev. E **64**, 025101R, (2001)
33. R. Albert, H. Jeong, A.-L. Barabási: Nature, **406**, 6794, 378 (2000)
34. R.V. Sole, J.M. Montoya: Proc. Roy. Soc. Lond. B Bio. **268**, 2039, (2001)
35. N. Schwartz, R. Cohen, D. ben-Avraham, A.-L. Barabasi, S. Havlin: Phys. Rev. E **66**, 015104(R) (2002)
36. P. Frojdh, M. Howard, K.B. Lauritsen: IJMPB **15**, 1761, (2001)
37. R. Pastor-Sattoras, A. Vespignani: Phys. Rev. E. **63**, 066117 (2001)
38. A.F. Rozenfed, R. Cohen, D. ben-Avraham, S. Havlin: Phys. Rev. Lett. **89**, 218701 (2002)
39. C.P. Warren, L.M. Sander, I.M. Sokolov: Phys. Rev. E **66**, 56105 (2002)
40. S.-H. Yook, H. Jeong, A.-L. Barabasi: PNAS **99**, 13382 (2002)
41. Note that (2.53) is analogous to Levy distribution, see e.g. M.F. Shlesinger and J. Klafter: Phys. Rev. Lett. 54, 2551 (1985)
42. D. ben-Avraham and S. Havlin: *Diffusion and Reactions in Fractals and Disordered Systems* (Cambridge University Press, 2000)
43. D. Stauffer and A. Aharony: *Introduction to Percolation Theory*, 2nd edition (Taylor and Francis, London, 1991)
44. S.N. Dorogovtsev, J.F.F. Mendes: Phys. Rev. E **63**, 062101 (2001)
45. We choose $m = 2d$ so that in the limit $\lambda \to \infty$ the network is identical with the embedding lattice. Clearly, this choice is not mandatory
46. Z. Burda, A. Krzywicki: cond-mat/0207020
47. S.S. Manna, P. Sen: cond-mat/0203216
48. R. Xulvi-Brunet, I.M. Sokolov: Phys. Rev. E **66**, 26118 (2002)

3 Hierarchical Organization of Modularity in Complex Networks

Albert-László Barabási[1], Erzsébet Ravasz[1], and Zoltán Oltvai[2]

[1] Department of Physics, 225 Nieuwland Science Hall, University of Notre Dame, Notre Dame, IN 46556, USA
[2] Department of Pathology, Northwestern University, Chicago, IL 60611, USA

Abstract. Many real networks in nature and society share two generic properties: they are scale-free and they display a high degree of clustering. We show that the scale-free nature and high clustering of real networks are the consequence of a hierarchical organization, implying that small groups of nodes form increasingly large groups in a hierarchical manner, while maintaining a scale-free topology. In hierarchical networks the clustering coefficient follows a strict scaling law, which can be used to identify the presence of a hierarchical organization in real networks. We find that several real networks, such as the World Wide Web, actor network, the Internet at the domain level and the semantic web obey this scaling law, indicating that hierarchy is a fundamental characteristic of many complex systems. We the focus on the metabolic network of 43 distinct organisms and show that many small, highly connected topologic modules combine in a hierarchical manner into larger, less cohesive units, their number and degree of clustering following a power law. Within *Escherichia Coli* we find that the uncovered hierarchical modularity closely overlaps with known metabolic functions.

3.1 Introduction

The availability of detailed network maps, capturing the topology of such diverse systems as the cell [1, 2, 3, 4], the world wide web [5], or the sexual network [6], have offered scientists for the first time the chance to address in quantitative terms the generic features of real networks (for reviews see [7, 8]). As a result, we learned that networks are governed by strict organizing principles, that generate systematic and measurable deviations from the topology predicted by the random graph theory of Erdős and Rényi [9, 10], the model used to describe complex webs in the past four decades.

Two properties of real networks have generated considerable attention. First, many networks display a high degree of clustering, measured by the clustering coefficient, which for node i with k_i links has the value $C_i = 2n_i/k_i(k_i-1)$, where n_i is the number of links between the k_i neighbors of i. Empirical results indicate that C_i averaged over all nodes is significantly higher for many real networks than for a random network of similar size [11, 7, 8]. Furthermore, the clustering coefficient of real networks is to a high degree independent of the number of nodes in the network (see Fig. 9 in [7]). At the same time, many networks of scientific or technological interest, ranging from the World Wide Web [5] to biological networks [1, 2, 3, 4] have been found to be scale-free [12, 13], which

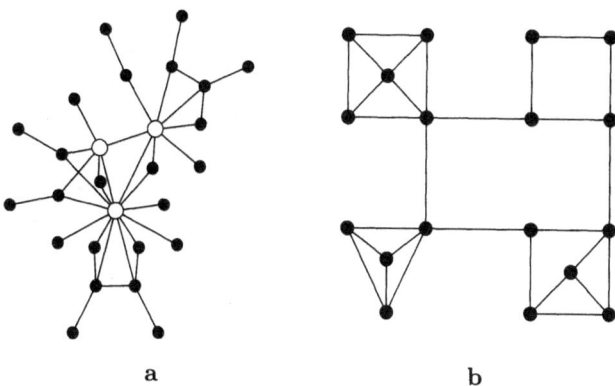

Fig. 3.1. a A schematic illustration of a scale-free network, whose degree distribution follows a power law. In such a network a few highly connected nodes, or hubs (*empty circles*) play an important role in keeping the whole network together. **b** Schematic illustration of a manifestly modular network made of four highly interlinked modules connected to each other by a few links. This intuitive topology does not have a scale-free degree distribution, as most of its nodes have a similar number of links, and hubs are absent (After [17])

means that the probability that a node has k links (i.e. degree k) follows

$$P(k) \sim k^{-\gamma} \, ,$$

where γ is the degree exponent.

The scale-free property and clustering are not exclusive: for a large number of real networks, including metabolic networks [1, 2], the protein interaction network [3, 4], the world wide web [5] and some social networks [14, 15, 16] the scale-free topology and high clustering coexist. Yet, most models proposed to describe the topology of complex networks have difficulty capturing simultaneously these two features. For example, the random network model [9, 10] cannot account neither for the scale-free, nor for the clustered nature of real networks, as it predicts an exponential degree distribution, and the average clustering coefficient, $C(N)$, decreases as N^{-1} with the number of nodes in the network. Scale-free networks (Fig. 3.1a), capturing the power law degree distribution, predict a much larger clustering coefficient than a random network. Indeed, numerical simulations indicate that for one of the simplest models [12, 13] the average clustering coefficient depends on the system size as $C(N) \sim N^{-0.75}$ [7, 8], significantly larger for large N than the random network prediction $C(N) \sim N^{-1}$. Yet, this prediction still disagrees with the finding that for several real systems C is independent of N [7].

On the biological front, it is now widely recognized that the thousands of components of a living cell are dynamically interconnected, so that the cell's functional properties are ultimately encoded into a complex intracellular web of molecular interactions [18, 19, 20, 21, 22, 23]. On the other hand, the identification and characterization of system-level features of biological organization is a key issue of post-genomic biology [24, 18, 19]. The concept of modularity assumes that cellular functionality can be seamlessly partitioned into a collection of modules. Each module is a discrete entity of several elementary components and performs an identifiable task, separable from the functions of other modules [24, 20, 21, 22, 25, 23]. Spatially and chemically isolated molecular machines or protein complexes (such as ribosomes and flagella) are prominent examples of such functional units, but more extended modules, such as those achieving their isolation through the initial binding of a signaling molecule [26] are also apparent.

The dilemma of modular versus highly integrated topology is perhaps most evident when inspecting cellular metabolism, a fully connected biochemical network in which hundreds of metabolic substrates are densely integrated via biochemical reactions. Within this network, however, modular organization (i.e., clear boundaries between sub-networks) is not immediately apparent.

A number of approaches for analyzing the functional capabilities of metabolic networks clearly indicate the existence of separable functional elements [27, 28]. Also, from a purely topologic perspective the metabolic network of *Escherichia coli* is known to possess a high clustering coefficient [2], a property that is suggestive of a modular organization. In itself, this implies that the metabolism of *E. coli* has a modular topology, potentially comprising several densely interconnected functional modules of varying sizes that are connected by few inter-module links (Fig. 3.1b). However, such clearcut modularity imposes severe restrictions on the degree distribution, implying that most nodes have approximately the same number of links, which contrasts with the metabolic network's scale-free nature [1, 2]. To determine if such a dichotomy is indeed a generic property of all metabolic networks we first calculated the average clustering coefficient for 43 different organisms [29] as a function of the number of distinct substrates, N, present in their metabolism. We find that for all 43 organisms the clustering coefficient is about an order of magnitude larger than that expected for a scale-free network of similar size (Fig. 3.2), suggesting that metabolic networks in all organisms are characterized by a high intrinsic potential modularity. We also observe that in contrast with the prediction of the scale-free model, for which the clustering coefficient decreases as $N^{-0.75}$ [7], the clustering coefficient of metabolic networks' is independent of their size (Fig. 3.2).

Here we show that the fundamental discrepancy between models and empirical measurements is rooted in a previously disregarded, yet generic feature of many real networks, biological and non-biological: their hierarchical topology. Indeed, in many networks one can easily identify groups of nodes that are highly interconnected with each other, but have only a few or no links to nodes outside of the group to which they belong to. In society such modules represent groups of friends or coworkers [30]; in the WWW denote communities with shared interests [31, 32]; in the actor network they characterize specific genres or simply individual movies. Some groups are small and tightly linked, others are larger and somewhat less interconnected. This clearly identifiable modular organiza-

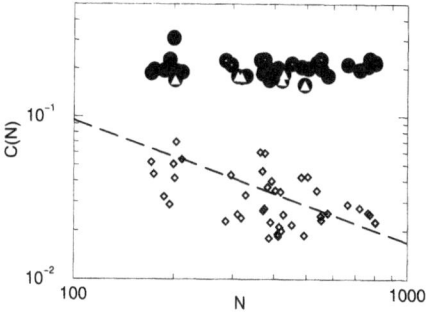

Fig. 3.2. The average clustering coefficient, $C(N)$, for 43 organisms [1] is shown as a function of the number of substrates N present in each of them. Species belonging to Archae (*white star*), Bacteria (*black circle*), and Eukaryotes (*white triangle*) are shown. The dashed line indicates the dependence of the clustering coefficient on the network size for a module-free scale-free network, while the diamonds denote C for a scale-free network with the same parameters (N and number of links) as observed in the 43 organisms (After [17])

tion is at the origin of the high clustering coefficient seen in many real networks. Yet, models reproducing the scale-free property of real networks [7, 8] distinguish nodes based only on their degree, and are blind to node characteristics that could lead to a modular topology.

In order to bring modularity, the high degree of clustering and the scale-free topology under a single roof, we need to assume that modules combine into each other in a hierarchical manner, generating what we call a *hierarchical network*. The presence of a hierarchy and the scale-free property impose strict restrictions on the number and the degree of cohesiveness of the different groups present in a network, which can be captured in a quantitative manner using a scaling law, describing the dependence of the clustering coefficient on the node degree. We use this scaling law to identify the presence of a hierarchical architecture in several real networks, and the absence of such hierarchy in geographically organized webs.

3.2 Hierarchical Network Model

We start by constructing a hierarchical network model, that combines the scale-free property with a high degree of clustering. Our starting point is a small cluster of five densely linked nodes (Fig. 3.3a). Next we generate four replicas of this hypothetical module and connect the four external nodes of the replicated clusters to the central node of the old cluster, obtaining a large 25–node module (Fig. 3.3b). Subsequently, we again generate four replicas of this 25–node module, and connect the 16 peripheral nodes to the central node of the old module (Fig. 3.3c), obtaining a new module of 125 nodes. These replication and connec-

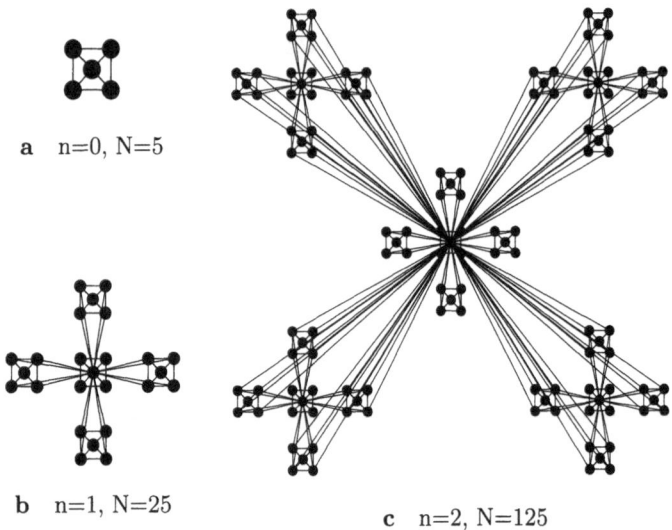

a   n=0, N=5

b   n=1, N=25

c   n=2, N=125

Fig. 3.3. The iterative construction leading to a hierarchical network. Starting from a fully connected cluster of five nodes shown in **a** (note that the diagonal nodes are also connected – links not visible), we create four identical replicas, connecting the peripheral nodes of each cluster to the central node of the original cluster, obtaining a network of $N = 25$ nodes **b**. In the next step we create four replicas of the obtained cluster, and connect the peripheral nodes again, as shown in **c**, to the central node of the original module, obtaining a $N = 125$ node network. This process can be continued indefinitely (After [33])

tion steps can be repeated indefinitely, in each step increasing the number of nodes in the system by a factor five.

Precursors to the model described in Fig. 3.3 have been proposed in [34] and extended and discussed in [35, 36] as a method of generating deterministic scale-free networks. Yet, it was believed that aside from their deterministic structure, their statistical properties are equivalent with the stochastic models that are often used to generate scale-free networks. In the following we argue that such hierarchical construction generates an architecture that is significantly different from the networks generated by traditional scale-free models. Most important, we show that the new feature of the model, its hierarchical character, are shared by a significant number of real networks.

First we note that the hierarchical network model seamlessly integrates a scale-free topology with an inherent modular structure. Indeed, the generated network has a power law degree distribution with degree exponent $\gamma = 1 + \ln 5/\ln 4 = 2.161$ (Fig. 3.4a). Furthermore, numerical simulations indicate that the clustering coefficient, $C \simeq 0.743$, is independent of the size of the network (Fig. 3.4c). Therefore, the high degree of clustering and the scale-free property are simultaneously present in this network.

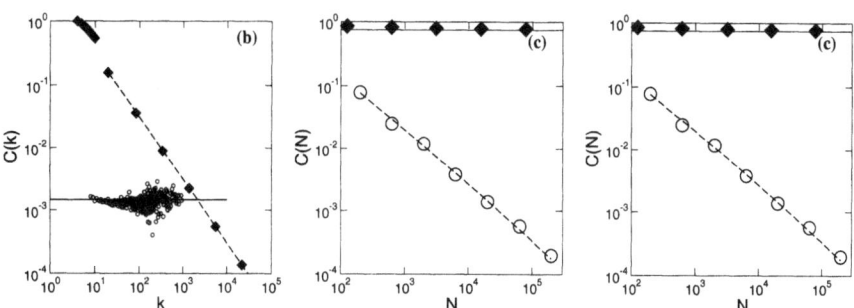

Fig. 3.4. Scaling properties of the hierarchical model shown in Fig. 3.3 ($N = 5^7$). **a** The numerically determined degree distribution. The assymptotic scaling, with slope $\gamma = 1 + \ln 5 / \ln 4$, is shown as a dashed line. **b** The $C(k)$ curve for the model, demonstrating that it follows (3.1). The open circles show $C(k)$ for a scale-free model [12] of the same size, illustrating that it does not have a hierarchical architecture. **c** The dependence of the clustering coefficient, C, on the size of the network N. While for the hierarchical model C is independent of N ($\blacklozenge$), for the scale-free model $C(N)$ decreases rapidly ($\circ$)

The most important feature of the network model of Fig. 3.3, not shared by either the scale-free [12, 13] or random network models [9, 10], is its hierarchical architecture. The network is made of numerous small, highly integrated five node modules (Fig. 3.3a), which are assembled into larger 25–node modules (Fig. 3.3b). These 25–node modules are less integrated but each of them is clearly separated from the other 25–node modules when we combine them into the even larger 125–node modules (Fig. 3.3c). These 125–node modules are even less cohesive, but again will appear separable from their replicas if the network expands further.

This intrinsic hierarchy can be characterized in a quantitative manner using the recent finding of Dorogovtsev, Goltsev and Mendes [35] that in deterministic scale-free networks the clustering coefficient of a node with k links follows the scaling law

$$C(k) \sim k^{-1} \ . \tag{3.1}$$

We argue that this scaling law quantifies the coexistence of a hierarchy of nodes with different degrees of clustering, and applies to the model of Fig. 3.3a–c as well. Indeed, the nodes at the center of the numerous 5–node modules have a clustering coefficient $C = 1$. Those at the center of a 25–node module have $k = 20$ and $C = 3/19$, while those at the center of the 125–node modules have $k = 84$ and $C = 3/83$, indicating that the higher a node's degree the smaller is its clustering coefficient, asymptotically following the $1/k$ law (Fig. 3.4b). In contrast, for the scale-free model proposed in [12] the clustering coefficient is independent of k, i.e. the scaling law (3.1) does not apply (Fig. 3.4b). The same is true for the random [9, 10] or the various small world models [11, 37], for which the clustering coefficient is independent of the nodes' degree.

Therefore, the discrete model of Fig. 3.3 combines within a single framework the two key properties of real networks: their scale-free topology and high modularity, which results in a system-size independent clustering coefficient. Yet, the hierarchical modularity of the model results in the scaling law (3.1), which is not shared by the traditional network models. The question is, could hierarchical modularity, as captured by this model, characterize real networks as well?

3.3 Hierarchical Organization in Non-biological Networks

To investigate if such hierarchical organization is present in real networks we measured the $C(k)$ function for several networks for which large topological maps are available. Next we discuss each of these systems separately.

Actor Network: Starting from the www.IMDB.com database, we connect any two actors in Hollywood if they acted in the same movie, obtaining a network of 392,340 nodes and 15,345,957 links. Earlier studies indicate that this network is scale-free with an exponential cutoff in $P(k)$ for high k [12, 38, 39]. As Fig. 3.5a

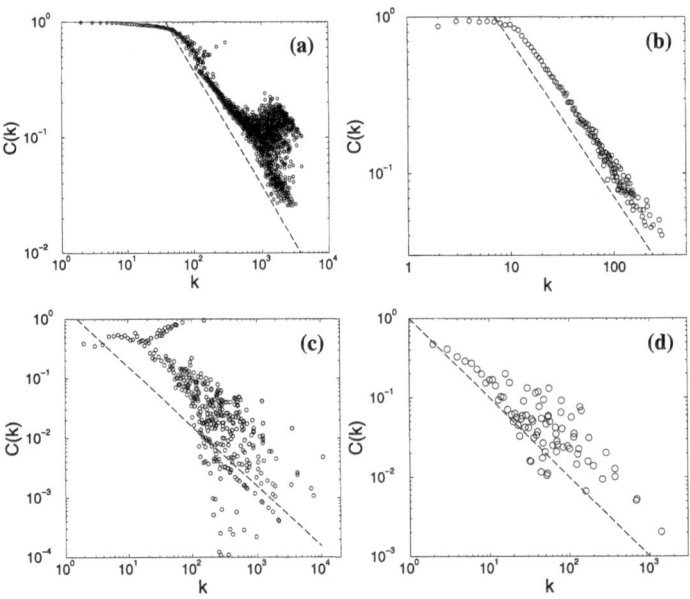

Fig. 3.5. The scaling of $C(k)$ with k for four large networks: **a** Actor network, two actors being connected if they acted in the same movie according to the www.IMDB.com database. **b** The semantic web, connecting two English words if they are listed as synonyms in the Merriam Webster dictionary [41]. **c** The World Wide Web, based on the data collected in [5]. **d** Internet at the Autonomous System level, each node representing a domain, connected if there is a communication link between them. The dashed line in each figure has slope -1, following (3.1) (After [33])

indicates, we find that $C(k)$ scales as k^{-1}, indicating that the network has a hierarchical topology. Indeed, the majority of actors with a few links (small k) appear only in one movie. Each such actor has a clustering coefficient equal to one, as all actors the actor has links to are part of the same cast, and are therefore connected to each other. The high k nodes include many actors that acted in several movies, and thus their neighbors are not necessarily linked to each other, resulting in a smaller $C(k)$. At high k the $C(k)$ curve splits into two branches, one of which continues to follow (3.1), while the other saturates. One explanation of this split is the decreasing amount of datapoints available in this region. Indeed, in the high k region the number of nodes having the same k is rather small. If one of these nodes corresponds to an actor that played only in a few movies with hundreds in the cast, it will have both high k and high C, considerably increasing the average value of $C(k)$. The k values for which such a high C nodes are absent continue to follow the k^{-1} curve, resulting in jumps between the high and small C values for large k. For small k these anomalies are averaged out.

Language network: Recently a series of empirical results have shown that the language, viewed as a network of words, has a scale-free topology [40, 41, 42, 43]. Here we study the network generated connecting two words to each other if they appear as synonyms in the Merriam Webster dictionary [41]. The obtained semantic web has 182,853 nodes and 317,658 links and it is scale-free with degree exponent $\gamma = 3.25$. The $C(k)$ curve for this language network is shown in Fig. 3.5b, indicating that it follows (3.1), suggesting that the language has a hierarchical organization.

World Wide Web: On the WWW two documents are connected to each other if there is an URL pointing from one document to the other one. The sample we study, obtained by mapping out the www.nd.edu domain [5], has 325,729 nodes and 1,497,135 links, and it is scale-free with degree exponents $\gamma_{\text{out}} = 2.45$ and $\gamma_{\text{in}} = 2.1$, characterising the out and in-degree distribution, respectively. To measure the $C(k)$ curve we made the network undirected. While the obtained $C(k)$, shown in Fig. 3.5c, does not follow as closely the scaling law (3.1) as observed in the previous two examples, there is clear evidence that $C(k)$ decreases rapidly with k, supporting the coexistence of many highly interconnected small nodes with a few larger nodes, which have a much lower clustering coefficient.

Indeed, the Web is full of groups of documents that all link to each other. For example, www.nd.edu/~networks, our network research dedicated site, has a high clustering coefficient, as the documents it links to have links to each other. The site is one of the several network-oriented sites, some of which point to each other. Therefore, the network research community still forms a relatively cohesive group, albeit less interconnected than the www.nd.edu/~networks site, thus having a smaller C. This network community is nested into the much larger community of documents devoted to statistical mechanics, that has an even smaller clustering coefficient. Therefore, the k–dependent $C(k)$ reflects the hierarchical nesting of the different interest groups present on the Web. Note that $C(k) \sim k^{-1}$ for the WWW was observed and briefly noted in [44].

Internet at the AS Level: The Internet is often studied at two different levels of resolution. At the router level we have a network of routers connected by various physical communication links. At the interdomain or autonomous system (AS) level each administrative domain, composed of potentially hundreds of routers, is represented by a single node. Two domains are connected if there is at least one router that connects them. Both the router and the domain level topology have been found to be scale-free [45]. As Fig. 3.5d shows, we find that at the domain level the Internet, consisting of 65,520 nodes and 24,412 links [46], has a hierarchical topology as $C(k)$ is well approximated with (3.1). The scaling of the clustering coefficient with k for the Internet was earlier noted by Vazquez, Pastor-Satorras and Vespignani (VPSV) [47, 48], who observed $C(k) \sim k^{-0.75}$. VPSV interpreted this finding, together with the observation that the average nearest-neighbor connectivity also follows a power-law with the node's degree, as a natural consequence of the *stub* and *transit* domains, that partition the network in a hierarchical fashion into international connections, national backbones, regional networks and local area networks.

Our measurements indicate, however, that some real networks lack a hierarchical architecture, and do not obey the scaling law (3.1). In particular, we find that the power grid and the router level Internet topology have a k independent $C(k)$.

Internet at the Router Level: The router level Internet has 260,657 nodes connected by 1,338,100 links [49]. Measurements indicate that the network is scale-free [45, 50] with degree exponent $\gamma = 2.23$. Yet, the $C(k)$ curve (Fig. 3.6a), apart from some fluctuations, is largely independent of k, in strong contrast with the $C(k)$ observed for the Internet's domain level topology (Fig. 3.5d), and in agreement with the results of VPSV [47, 48], who also note the absence of a hierarchy in router level maps.

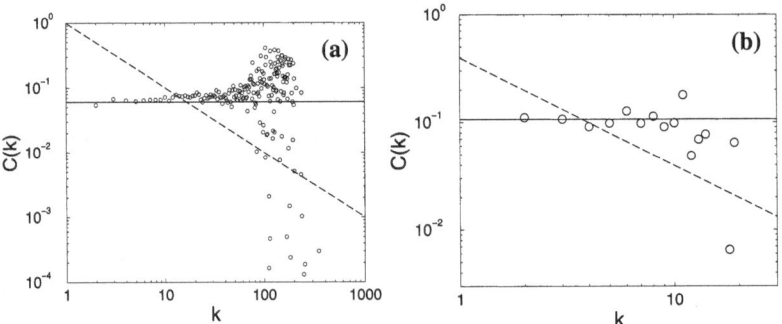

Fig. 3.6. The scaling of $C(k)$ for two large, non-hierarchical networks: **a** Internet at router level [49]. **b** The power grid of Western United States. The dashed line in each figure has slope -1, while the solid line corresponds to the average clustering coefficient (After [33])

Power Grid: The nodes of the power grid are generators, transformers and substations and the links are high voltage transmission lines. The network studied by us represents the map of the Western United States, and has 4,941 nodes and 13,188 links [11]. The results again indicate that apart from fluctuations, $C(k)$ is independent of k.

It is quite remarkable that these two networks share a common feature: a geographic organization. The routers of the Internet and the nodes of the power grid have a well defined spatial location, and the link between them represent physical links. In contrast, for the examples discussed in Fig. 3.5 the physical location of the nodes was either undefined or irrelevant, and the length of the link was not of major importance. For the router level Internet and the power grid the further are two nodes from each other, the more expensive it is to connect them [50]. Therefore, in both systems the links are driven by cost considerations, generating a distance driven structure, apparently excluding the emergence of a hierarchical topology. In contrast, the domain level Internet is less distance driven, as many domains, such as the AT&T domain, span the whole United States.

In summary, we offered evidence that for four large networks $C(k)$ is well approximated by $C(k) \sim k^{-1}$, in contrast to the k–independent $C(k)$ predicted by both the scale-free and random networks. In addition, there is evidence for similar scaling in the metabolism [17] and protein interaction networks [51]. This indicates that these networks have an inherently hierarchical organization. In contrast, hierarchy is absent in networks with strong geographical constraints, as the limitation on the link length strongly constrains the network topology.

3.4 Hierarchy in Metabolic Networks and the Functional Organization of *Escherichia Coli*

To investigate if hierarchical organization is present in cellular metabolism we measured the $C(k)$ function for the metabolic networks of all 43 organisms. As shown in Fig. 3.7, for each organism $C(k)$ is well approximated by $C(k) \sim k^{-1}$, in contrast to the k–independent $C(k)$ predicted by both the scale-free and modular networks. This provides direct evidence for an inherently hierarchical organization.

A key issue from a biological perspective is whether the identified hierarchical architecture reflects the true functional organization of cellular metabolism. To uncover potential relationships between topological modularity and the functional classification of different metabolites we concentrate on the metabolic network of *Escherichia coli*, whose metabolic reactions have been exhaustively studied, both biochemically and genetically [52].

Using a previously established graph-theoretical representation [1], we first subjected *E. coli*'s metabolic organization to a three step reduction process, replacing non-branching pathways with equivalent links, allowing us to decrease its complexity without altering the network topology [54]. Next, we calculated

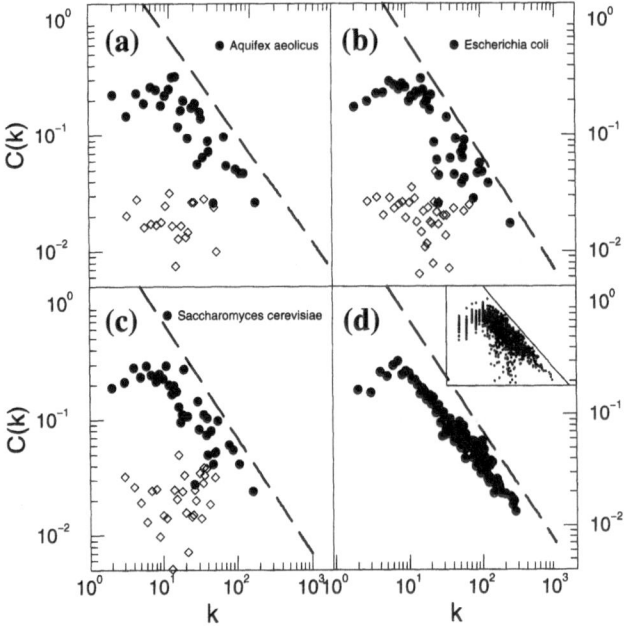

Fig. 3.7. The dependence of the clustering coefficient on the node's degree in three organisms: **a** *Aquidex Aeolicus* (archaea), **b** *Escherichia Coli* (bacterium), **c** and *Saccharomices cerevisiae* (eukaryote). In d the $C(k)$ curves averaged over all 43 organisms are shown, while the inset displays all 43 species together. The dashed lines correspond to $C(k) \sim k^{-1}$, and in **a–c** the diamonds represent $C(k)$ expected for a scale-free network (Fig. 3.1a) of similar size, indicating the absence of scaling. The wide fluctuations are due to the small size of the network (After [17])

the topological overlap matrix, $O_{\mathrm{T}}(i,j)$, of the condensed metabolic network (Fig. 3.8). A topological overlap of one between substrates i and j implies that they are connected to the same substrates, while a zero value indicates that i and j do not share links to common substrates among the metabolites they react with.

The metabolites that are part of highly integrated modules have a high topological overlap with their neighbors, and we find that the larger the overlap between two substrates within the *E. coli* metabolic network the more likely it is that they belong to the same functional class.

As the topological overlap matrix is expected to encode the comprehensive functional relatedness of the substrates forming the metabolic network, we investigated whether potential functional modules encoded in the network topology can be uncovered automatically. Initial application of an average-linkage hierarchical clustering algorithm [53] to the overlap matrix of the small hypothetical network shown in Fig. 3.8a placed those nodes that have a high topological overlap close to each other (Fig. 3.8b). Also, the method has clearly identified the

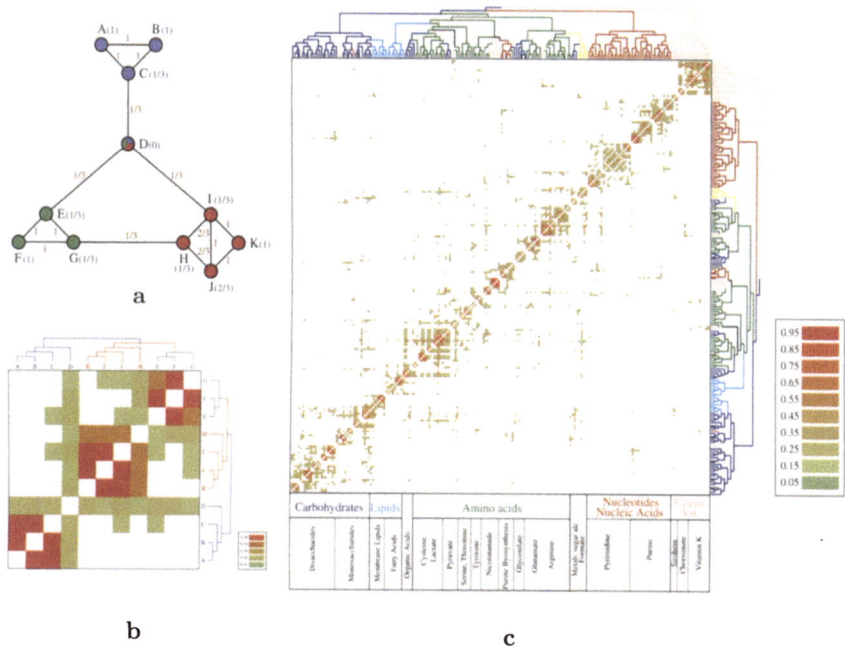

Fig. 3.8. a Topological overlap illustrated on a small hypothetical network. For each pair of nodes, i and j, we define the topological overlap $O_T(i,j) = J_n(i,j)/[\min(k_i, k_j) + 1 - L(i,j)]$, where $J_n(i,j)$ denotes the number of nodes to which both i and j are linked to plus $L(i,j)$, which is one if there is a direct link between i and j, zero otherwise, and $\min(k_i, k_j)$ is the smaller of the k_i and k_j degrees. On each link we indicate the topological overlap for the connected nodes and in parenthesis next to each node we indicate it's clustering coefficient. **b** The topological overlap matrix corresponding to the small network shown in **a**. The rows and columns of the matrix were reordered by the application of an average linkage clustering method [53] to its elements, allowing us to identify and place close to each other those nodes that have high topological overlap. The color code denotes the degree of topological overlap between the nodes (see sidebar). The associated tree clearly reflects the three distinct modules built into the model of **a**, as well as the fact that the EFG and HIJK modules are closer to each other in topological sense that the ABC module. **c** The topologic overlap matrix corresponding to the *E. coli* metabolism, together with the corresponding hierarchical tree (*top*) that quantifies the relationship between the different modules. The branches of the tree are color coded to reflect the functional classification of their substrates. The biochemical classes we used to group the metabolites represent carbohydrate metabolism (*blue*), nucleotide and nucleic acid metabolism (*red*), protein, peptide and amino acid metabolism (*green*), lipid metabolism (*cyan*), aromatic compound metabolism (*dark pink*), monocarbon compound metabolism (*yellow*) and coenzyme metabolism (*light orange*) [29]. The color code of the matrix denotes the degree of topological overlap shown in the matrix. On the bottom we show the large-scale functional map of the metabolism, as suggested by the hierarchical tree (After [17])

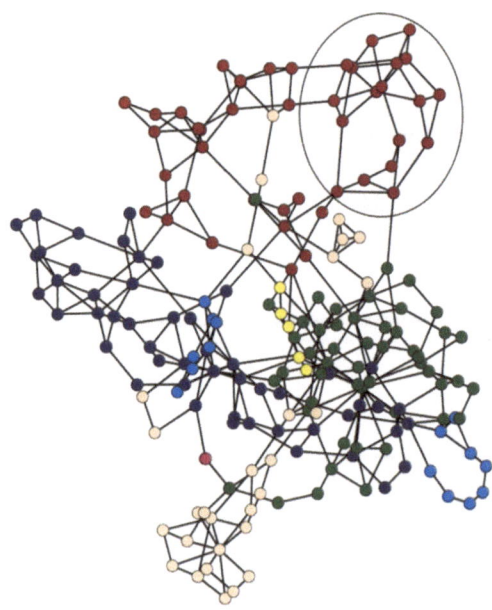

Fig. 3.9. 3-D representation of the reduced *E. coli* metabolic network. Each node is color coded by the functional class to which it belongs, and is identical to the color code applied to the branches of the tree shown in Fig. 3.8c. Note that the different functional classes are visibly segregated into topologically distinct regions of metabolism. The blue-shaded region denotes the nodes belonging to pyrimidine metabolism, discussed below (After [17])

three distinct modules built into the model of Fig. 3.8a, as illustrated by the fact that the EFG and HIJK modules are closer to each other in a topological sense than the ABC module (Fig. 3.8b).

Application of the same technique on the *E. coli* overlap matrix $O_T(i, j)$ provides a global topologic representation of *E. coli* metabolism (Fig. 3.8c). Groups of metabolites forming tightly interconnected clusters are visually apparent, and upon closer inspection the hierarchy of nested topologic modules of increasing sizes and decreasing interconnectedness are also evident. To visualize the relationship between topological modules and the known functional properties of the metabolites, we color coded the branches of the derived hierarchical tree according to the predominant biochemical class of the substrates it produces, using the standard, small molecule biochemistry based classification of metabolism [29].

As shown in Fig. 3.8c, and in the three dimensional representation in Fig. 3.9, we find that most substrates of a given small molecule class are distributed on the same branch of the tree (Fig. 3.8c) and correspond to relatively well-delimited regions of the metabolic network (Fig. 3.9). Therefore, there are strong correlations between shared biochemical classification of metabolites and the global topological organization of *E. coli* metabolism (Fig. 3.8c, bottom, and [54]).

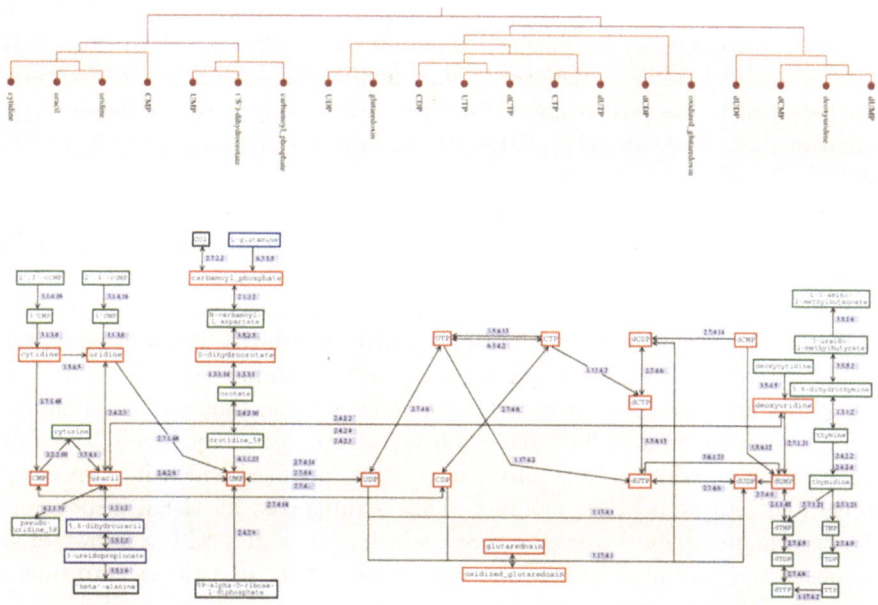

Fig. 3.10. A detailed diagram of the metabolic reactions that surround and incorporate the pyrimidine metabolic module. Red boxes denote the substrates directly appearing in the reduced metabolism and the tree shown in Fig. 3.9. Substrates in green boxes are internal to pyrimidine metabolism, but represent members of non-branching pathways or end pathways branching from a metabolite with multiple connections [54]. Blue and black boxes show the connections of pyrimidine metabolites to other parts of the metabolic network. Black boxes denote core substrates belonging to other branches of the metabolic tree Fig. 3.8c, while blue boxes denote non-branching pathways (if present) leading to those substrates. The shaded boxes around the reactions highlight the modules suggested by the hierarchical tree. The shaded blue boxes along the links display the enzymes catalyzing the corresponding reactions, and the arrows show the direction of the reactions according to the WIT metabolic maps [29]. (After [17])

To correlate the putative modules obtained from our graph theory-based analysis to actual biochemical pathways, we concentrated on the pathways involving the pyrimidine metabolites. Our method divided these pathways into four putative modules (Fig. 3.10a), which represent a topologically well-limited area of *E. coli* metabolism (Fig. 3.9, *circle*).

As shown in Fig. 3.10b, all highly connected metabolites (Fig. 3.10b, *red boxes*) correspond to their respective biochemical reactions within pyrimidine metabolism, together with those substrates that were removed during the original network reduction procedure, and then re-added (Fig. 3.10b, *green boxes*). However, it is also apparent that putative module boundaries do not always overlap with intuitive 'biochemistry-based' boundaries. For instance, while the synthesis of UMP from L-glutamine is expected to fall within a single module based on a linear set of biochemical reactions, the synthesis of UDP from UMP leaps putative module boundaries. Thus, further experimental and theoretical analyses will be needed to understand the relationship between the decomposition of *E. coli* metabolism offered by our topology-based approach, and the biologically relevant sub-networks.

The organization of metabolic networks is likely to combine a capacity for rapid flux reorganization with a dynamic integration with all other cellular function [2]. Our results indicate that the system-level structure of cellular metabolism is best approximated by a hierarchical network organization with seamlessly embedded modularity. In contrast to current, intuitive views of modularity (Fig. 3.1b) which assume the existence of a set of modules with a non-uniform size potentially separated from other modules, we find that the metabolic network has an inherent self-similar property: there are many highly integrated small modules, which group into a few larger modules, which in turn can be integrated into even larger modules. This is supported by visual inspection of the derived hierarchical tree (Fig. 3.8c), which offers a natural breakdown of metabolism into several large modules, which are further partitioned into smaller, but more integrated sub-modules.

3.5 Stochastic Model and Universality

The hierarchical model described in Fig. 3.3 predicts $C(k) \sim k^{-1}$, which offers a rather good fit to three of the four $C(k)$ curves shown in Fig. 3.5. The question is, is this scaling law (3.1) universal, valid for all hierarchical networks, or could different scaling exponent characterize the scaling of $C(k)$? Defining the hierarchical exponent, β, as

$$C(k) \sim k^{-\beta} , \tag{3.2}$$

is $\beta = 1$ a universal exponent, or it's value can be changed together with γ? In the following we demonstrate that the hierarchical exponent β can be tuned as we tune some of the network parameters. For this we propose a stochastic version of the model described in Fig. 3.3.

We start again with a small core of five nodes all connected to each other (Fig. 3.3a) and in step one ($n = 1$) we make four copies of the five node module. Next, we randomly pick a p fraction of the newly added nodes and connect each of them independently to the nodes belonging to the central module. We use preferential attachment [12, 13] to decide to which central node the selected nodes link to. That is, we assume that the probability that a selected node will connect to a node i of the central module is $k_i / \sum_j k_j$, where k_i is the degree of node i and the sum goes over all nodes of the central module. In the second step ($n = 2$) we again create four identical copies of the 25–node structure obtained thus far, but we connect only a p^2 fraction of the newly added nodes to the

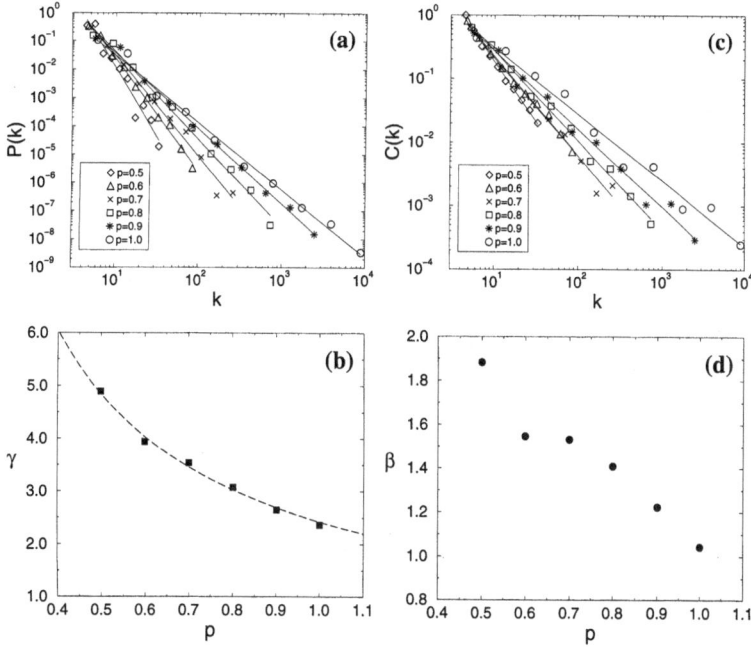

Fig. 3.11. The scaling properties of the stochastic model. **a** The degree distribution for different p values, indicating that $P(k)$ follows a power law with a p dependent slope. **b** The dependence of the degree exponent γ on p, determined by fitting power laws to the curves shown in **a**. The exponent γ appears to follow approximately $\gamma(p) \sim 1/p$ (dashed line). **c** The $C(k)$ curve for different p values, indicating that the hierarchical exponent β depends on p. **d** The dependence of β on the parameter p. The simulations were performed for $N = 5^7 (78,125)$ nodes (After [33])

central module. Subsequently, in each iteration n the central module of size 5^n is replicated four times, and in each new module a p^n fraction will connect to the current central module, requiring the addition of $(5p)^n$ new links.

As Fig. 3.11 shows, changing p alters the slope of both $P(k)$ and $C(k)$ on a log–log plot. In general, we find that increasing p decreases the exponents γ and β (Fig. 3.11b,d). The exponent $\beta = 1$ is recovered for $p = 1$, i.e. when all nodes of a module gain a link. While the number of links added to the network changes at each iteration, for any $p \leq 1$ the average degree of the infinitely large network is finite. Indeed, the average degree follows

$$\langle k \rangle_n = \frac{8}{5} \left(\frac{3}{2} + \frac{1 - p^{n+1}}{1 - p} \right), \qquad (3.3)$$

which is finite for any $p \leq 1$.

Interestingly, the scaling of $C(k)$ is not a unique property of the model discussed above. A version of the model, where we keep the fraction of selected nodes, p, constant from iteration to iteration, also generates p dependent β and

γ exponents. Furthermore, recently several results indicate that the scaling of $C(k)$ is an intrinsic feature of several existing growing networks models. Indeed, aiming to explain the potential origin of the scaling in $C(k)$ observed for the Internet, VSPV note that the fitness model [55, 56] displays a $C(k)$ that appears to scale with k. While there is no analytical evidence for $C(k) \sim k^{-\beta}$ yet, numerical results [47, 48] suggest that the presence of fitness does generate a hierarchical network architecture. In contrast, in a recent model proposed by Klemm and Eguiluz there is analytical evidence that the network obeys the scaling law (3.1) [57]. In their model in each time step a new node joins the network, connecting to all *active* nodes in the system. At the same time an active node is deactivated with probability $p \sim k^{-1}$. The insights offered by the hierarchical model can help understand the origin of the observed $C(k) \sim k^{-1}$. By deactivating the less connected nodes a central core emerges to which all subsequent nodes tend to link to. New nodes have a large C and small k, thus they are rapidly deactivated, freezing into a large C state. The older, more connected, surviving nodes are in contact with a large number of nodes that have already disappeared from the active list, and they have small C^3.

Finally, Szabó, Alava and Kertész have developed a rate equation method to systematically calculate $C(k)$ for evolving networks models [59]. Applying the method to a model proposed by Holme and Kim [60] to enhance the degree of clustering coefficient C seen in the scale-free model [12], they have shown that the scaling of $C(k)$ depends on the parameter p, which governs the rate at which new nodes connect to the neighbors of selected nodes, bypassing preferential attachment. As for $p = 0$ the Holme–Kim model reduces to the scale-free model, Szabó, Alava and Kertész find that in this limit the scaling of $C(k)$ vanishes. These models indicate that several microscopic mechanisms could generate a hierarchical topology, just as several models are able to create a scale-free network [7, 8].

3.6 Discussion and Outlook

The identified hierarchical architecture offers a new perspective on the topology of complex networks. Indeed, the fact that many large networks are scale-free is now well established. It is also clear that most networks have a modular topology, quantified by the high clustering coefficient they display. Such modules have been proposed to be a fundamental feature of biological systems [24, 17], but have been discussed in the context of the WWW [61, 31], and social networks as well [30, 62]. The hierarchical topology offers a new avenue for bringing under a single roof these two concepts, giving a precise and quantitative meaning for the network's modularity. It indicates that we should not think of modularity as the

---

[3] Note, however, that as new nodes tend to connect to nodes that were added to the network shortly before them, the model generates a close to one dimensional structure in time. See e.g. [58]

coexistence of relatively independent groups of nodes. Instead, we have many small clusters, that are densely interconnected. These combine to form larger, but less cohesive groups, which combine again to form even larger and even less interconnected clusters. This self-similar nesting of different groups or modules into each other forces a strict fine structure on real networks.

For biological systems hierarchical modularity is consistent with the notion that evolution may act at many organizational levels simultaneously: the accumulation of many local changes, that affect the small, highly integrated modules, could slowly impact the properties of the larger, less integrated modules. The emergence of the hierarchical topology via copying and reusing existing modules [24] and motifs [23], a process reminiscent of the results of gene duplication [63, 64], offers a special role to the modules that appeared first in the network. While the model of Fig. 3.4 reproduces the large-scale features of the metabolism, understanding the evolutionary mechanism that explains the simultaneous emergence of the observed hierarchical and scale-free topology of the metabolism, and its generality to cellular organization, is now a prime challenge.

Most interesting is, however, the fact that the hierarchical nature of these networks is well captured by a simple quantity, the $C(k)$ curve, offering us a relatively straightforward method to identify the presence of hierarchy in real networks. The law (3.1) indicates that the number and the size of the groups of different cohesiveness is not random, but follow rather strict scaling laws.

The presence of such a hierarchical architecture reinterprets the role of the hubs in complex networks. Hubs, the highly connected nodes at the tail of the power law degree distribution, are known to play a key role in keeping complex networks together, playing a crucial role from the robustness of the network [65, 66] to the spread of viruses in scale-free networks [67]. Our measurements indicate that the clustering coefficient characterizing the hubs decreases linearly with the degree. This implies that while the small nodes are part of highly cohesive, densely interlinked clusters, the hubs are not, as their neighbors have a small chance of linking to each other. Therefore, the hubs play the important role of bridging the many small communities of clusters into a single, integrated network.

In many ways our study offers only a starting point for understanding the interplay between the scale-free, hierarchical and modular nature of real networks. While the $C(k)$ curves offer a tool to unearth the presence of a hierarchy, it is unclear what are the minimal ingredients at the model level for such a hierarchy to emerge. Finally, the role of the geometrical factor, which appears to remove the hierarchy, needs to be elucidated. Further modeling and empirical studies should allow us to address these questions.

References

1. H. Jeong, B. Tombor, R. Albert, Z.N. Oltvai, A.-L. Barabási: Nature **407**, 651 (2000)
2. A. Wagner, D.A. Fell: Proc. R. Soc. Lond. B. Biol. Sci. **268**, 1803 (2001)
3. H. Jeong, S. Mason, A.-L. Barabási, Z.N. Oltvai: Nature **411**, 41 (2001)
4. A. Wagner: Mol. Biol. Evol. **18**, 1283 (2001)
5. R. Albert, H. Jeong, A.-L. Barabási: Nature **401**, 130 (1999)
6. F. Liljeros, C.R. Edling, L.A.N. Amaral, H.E. Stanley, Y. Åberg: Nature **411**, 907 (2001)
7. R. Albert, A.-L. Barabási: Rev. Mod. Phys. **74**, 47 (2002)
8. S.N. Dorogovtsev, J.F.F. Mendes: Adv. Phys. **51**, 1079 (2002)
9. P. Erdős, A. Rényi: Publ. Math. Debrecen **6**, 290 (1959)
10. B. Bollobás: *Random Graphs* (Academic Press, London 1985)
11. D.J. Watts, S.H. Strogatz: Nature **393**, 440 (1998)
12. A.-L. Barabási, R. Albert: Science **286**, 509 (1999)
13. A.-L. Barabási, R. Albert, H. Jeong: Physica A **272**, 173 (1999)
14. M.E.J. Newman: Proc. Nat. Acad. Sci. U.S.A **98**, 404 (2001)
15. M.E.J. Newman: Phys. Rev. E **64**, 016131 (2001)
16. A.-L. Barabási, H. Jeong, Z. Néda, E. Ravasz, A. Schubert, T. Vicsek: Physica A **311**, 590 (2002)
17. E. Ravasz, A.L. Somera, D.A. Mongru, Z.N. Oltvai, A.-L. Barabási: Science **297**, 1551 (2002)
18. H. Kitano: Science **295**, 1662 (2002)
19. Y.I. Wolf, G. Karev, E. V. Koonin: Bioessays **24**, 105 (2002)
20. D.A. Lauffenburger: Proc. Nact. Acad. Sci. U.S.A **97**, 5031 (2000)
21. C.V. Rao, A.P. Arkin: Annu. Rev. Biomed. Eng. **3**, 391 (2001)
22. N.S. Holter, A. Maritan, M. Cieplak, N.V. Feoroff, J. R. Banavar: Proc. Nact. Acad. Sci. U.S.A **98**, 1693 (2001)
23. S.S. Shen-Orr, R. Milo, S. Mangan, U. Alon: Nature Genet. **31** 64 (2002)
24. L.H. Hartwell, J.J. Hopfield, S. Leibler, A.W. Murray: Nature **402**, C47 (1999)
25. J. Hasty, D. McMillen, F. Isaacs, J.J. Collins: Nature Rev. Genet. **2**, 268 (2001)
26. U. Alon, M.G. Surette, N. Barkai, S. Leibler: Nature **397**, 168 (1999)
27. C.H. Schilling, D. Letscher, B.O. Palsson: J. Theor. Biol. **203**, 229 (2000)
28. S. Schuster, D.A. Fell, T. Dandekar: Nature Biotechnol. **18**, 326 (2000)
29. R. Overbeek et. al: Nucleic Acids Res. **28**, 123 (2000)
30. M.S. Granovetter: Am. J. Sociol. **78**, 1360 (1973)
31. G.W. Flake, S. Lawrence, C.L. Giles. In: *Proceedings of the Sixth International Conference on Knowledge Discovery and Data Mining, August, 2000* (ACM, Boston) pp. 150
32. L.A. Adamic, E. Adar: *Friends and neighbors on the web*, preprint available at http://hpl.hp.com/shl/papers/web10/index.html (2000)
33. E. Ravasz, A.-L. Barabási: Phys. Rev. E, in press (2002)
34. A.-L. Barabási, E. Ravasz, T. Vicsek: Physica A **299**, 559 (2001)
35. S.N. Dorogovtsev, A.V. Goltsev, J.F.F. Mendes: Phys. Rev. E **65**, 066122 (2002)
36. S. Jung, S. Kim, B. Kahng: Phys. Rev. E **65**, 056101 (2002)
37. M.E.J. Newman: J. Stat. Phys. **101**, 819 (2000)
38. R. Albert, A.-L. Barabási: Phys. Rev. Lett. **85**, 5234 (2000)
39. L.A.N. Amaral, A. Scala, M. Barthélémy, H.E. Stanley: Proc. Nact. Acad. Sci. U.S.A **97**, 11149 (2000)

40. R. Ferrer i Cancho, R.V. Solé: Proc. R. Soc. Lond. B **268**, 2261 (2001)
41. S. Yook, H. Jeong, A.-L. Barabási: to be published (2002)
42. M. Sigman, G. Cecchi: Proc. Nac. Acad. Sci. U.S.A **99**, 1742 (2002)
43. S.N. Dorogovtsev, J.F.F. Mendes: Proc. R. Soc. Lond. B. **268**, 2603 (2001)
44. J.-P. Eckmann, E. Moses: Proc. Nact. Acad. Sci. U.S.A **99**, 5825 (2002)
45. M. Faloutsos, P. Faloutsos, C. Faloutsos: Comput. Commun. Rev. **29**, 251 (1999)
46. Data available at http://moat.nlanr.net/infrastructure.html
47. A.Vázquez, R. Pastor-Satorras, A. Vespignani: Phys. Rev. E **65**, 066130 (2002)
48. A.Vázquez, R. Pastor-Satorras, A. Vespignani: *Internet topology at the router and autonomous system level*, Los Alamos Archive cond-mat/0206084 (2002)
49. R. Govindan, H. Tangmunarunkit. In: *Proceedings of IEEE INFOCOM 2000, Tel Aviv, Israel, Vol. 3* (IEEE, Piscataway N. J. 2000) pp. 1371
50. S.H. Yook, H. Jeong, A.-L. Barabási: Proc. Nact. Acad. Sci. U.S.A **99**, 13382 (2002)
51. S.H. Yook, Z.N. Oltvai, A.-L. Barabási: submitted (2002)
52. P.D. Karp, M. Riley, S.M. Paley, A. Pellegrini-Toole, M. Krummenacker: Nucleic Acids Res. **30**, 56 (2002)
53. M.B. Eisen, P.T. Spellman, P.O. Brown, D. Botstein: Proc. Nact. Acad. Sci. U.S.A **95**, 14863 (1998)
54. Additional information is available at www.nd.edu/~ networks/cell/index.html.
55. G. Bianconi, A.-L. Barabási: Europhys. Lett. **54** (4), 436 (2001)
56. G. Bianconi, A.-L. Barabási: Phys. Rev. Lett. **86**, 5632 (2001)
57. K. Klemm, V.M. Eguiluz: Phys. Rev. E **65**, 036123 (2002)
58. A. Vázquez, Y. Moreno, M. Boguñá, R. Pastor-Satorras, A. Vespignani: *Topology and correlations in structured scale-free networks*, preprint (2002).
59. G. Szabó, M. Alava, J. Kertész: *Structural transitions in scale-free networks*, Los Alamos Archive cond-mat/0208551 (2002)
60. P. Holme, B.J. Kim: Phys. Rev. E **65**, 026107 (2002)
61. S. Lawrence, C.L. Giles: Nature **400**, 107 (1999)
62. D.J. Watts, P.S. Dodds, M.E.J. Newman: Science **296**, 1302 (2002)
63. A. Vásquez, A. Flamini, A. Martian, A. Vespignani: Phys. Rev. E **65**, 066130 (2002)
64. R.V. Solé, R. Pastor-Satorras, E.D. Smith, T. Kepler: Santa Fe Institute Working Paper 01-08-041   available at http://www.santafe.edu/sfi/publications/wpabstract/200108041
65. R. Albert, H. Jeong, A.-L. Barabási: Nature **406**, 378 (2000)
66. R. Cohen, K. Erez, D. ben Avraham, S. Havlin: Phys. Rev. Lett. **86**, 3682 (2001)
67. R. Pastor-Satorras, A. Vespignani: Phys. Rev. Lett. **86**, 3200 (2001)

4 Mixing Patterns and Community Structure in Networks

M.E.J. Newman[1,3] and M. Girvan[2,3]

[1] Department of Physics, University of Michigan, Ann Arbor, MI 48109, USA
[2] Department of Physics, Cornell University, Ithaca, NY 14853, USA
[3] Santa Fe Institute, 1399 Hyde Park Road, Santa Fe, NM 87501, USA

Abstract. Common experience suggests that many networks might possess community structure – division of vertices into groups, with a higher density of edges within groups than between them. Here we describe a new computer algorithm that detects structure of this kind. We apply the algorithm to a number of real-world networks and show that they do indeed possess non-trivial community structure. We suggest a possible explanation for this structure in the mechanism of assortative mixing, which is the preferential association of network vertices with others that are like them in some way. We show by simulation that this mechanism can indeed account for community structure. We also look in detail at one particular example of assortative mixing, namely mixing by vertex degree, in which vertices with similar degree prefer to be connected to one another. We propose a measure for mixing of this type which we apply to a variety of networks, and also discuss the implications for network structure and the formation of a giant component in assortatively mixed networks.

4.1 Introduction

Much of the recent research on the structure of networks of various kinds has looked at properties like path lengths, transitivity, degree distributions, and resilience of networks to vertex deletion [42, 2, 15], all of which, while of exceptional importance in many contexts, tend to focus our attention on the properties of individual vertices or vertex pairs – how far apart they are, what their degrees are, and so forth. However, in other contexts it may be equally important to ask about the large-scale properties of the network as a whole. Numbers of components and their distribution of sizes would be an example of such a property, one which is relevant to issues of accessibility [10] and to epidemiology [18, 7, 31]. Searchability and the performance of search algorithms on networks would be another [25, 1, 45]. A third is the existence and effects of large-scale inhomogeneity in networks – what we call "community structure", the presence (or absence) in the network of regions with high densities of connections between vertices and other regions with low densities – and it is with a discussion of this topic that we begin this paper. (In some circles, this phenomenon is called "clustering", an unfortunate terminology which risks confusion with another use of the word clustering introduced recently by Watts and Strogatz [46]. We will use the word clustering only in reference to hierarchical clustering, which is a standard technique for community detection; otherwise we will avoid it.)

Our investigation of community structure will lead us to consideration of mixing patterns in networks – which vertices connect to which others and why – as an explanation for observed communities in networks of all kinds, and eventually to consideration of more general classes of correlated networks including networks with correlations between the degrees of adjacent vertices.

Much of the work reported in this article has appeared previously in various papers, which the reader may like to consult for more detail than we can give here [17, 35, 36].

4.2 Community Structure

The oldest studies by far of the large-scale statistical properties of networks are the studies of social networks carried out within the sociological community, which stretch back at least to the 1930s [44, 41]. Social networks are network representations of relationships of some kind, generically called "ties", between people or groups of people, generically called "actors". Actors might be individuals, organizations or companies, while ties might represent friendship, acquaintance, business relationships or financial transactions, amongst other things.

A long-standing goal among social network analysts has been to find ways of analysing network data to reveal the structure of the underlying communities that they represent. It is commonly supposed that the actors in most social networks divide themselves naturally into groups of some kind, such that the density of ties within groups is higher than the density of ties between them. A sketch of a network with such community structure is shown in Fig. 4.1.

It is a matter of common experience that social networks do contain communities. We look around ourselves and see that we belong to this clique or that, that we have a circle of close friends and others whom we know less well, that there are groupings within our personal networks on the basis of interest, occu-

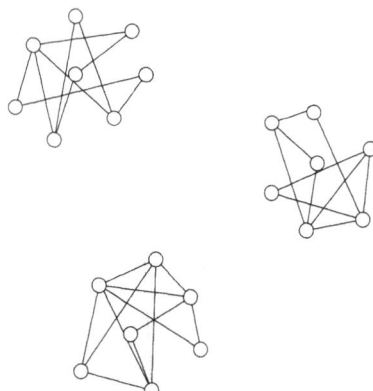

Fig. 4.1. A figurative sketch of a network possessing community structure of the type discussed here

pation, geographical location and so forth. This does not however guarantee that a network contains community structure of type that we are considering here. It would be perfectly possible for each person in a network to have a well-defined set of close acquaintances, their own personal network neighbourhood, but for the network neighbourhoods of different people to overlap only partially, so that the network as a whole is quite homogeneous, with no clear communities emerging from the pattern of vertices and edges. A network model showing precisely this type of structure has been proposed and studied recently by Kleinberg [26]. Our purpose in this section will be to investigate methods for detecting whether true community structure does exist in networks and for extracting the communities, and to apply those methods to particular networks. As we will see, the early intuition of the sociologists was correct, and many of the networks studied, including non-social networks, do possess large-scale inhomogeneity of precisely the type that would indicate the presence of community divisions.

The problem then is to take a network, specified in the simplest case by a list of n vertices joined in pairs by m edges, and from this structure to extract a set of communities – non-overlapping subsets of vertices that are, in some sense, tightly knit, having stronger within-group connections than between-group connections. The traditional, and still most common, method for detecting structure of this kind is the method of "hierarchical clustering" [44, 41]. In this method one defines a connection strength for each pair of vertices in the network, i.e., $\frac{1}{2}n(n-1)$ numbers that represent a distance or weight for the connection between each pair. (In some versions of the method not all pairs are assigned a connection strength, in which case those that are not can be assumed to have a connection strength of zero.) Examples of possible definitions for the strengths include geodesic (shortest path) distances between pairs, or their inverses if one wants a measure that increases when pairs are more closely connected, counts of numbers of vertex- or edge-independent paths between pairs ("maxflow" methods) or weighted counts of total numbers of paths between pairs (adjacency matrix methods).

Then, starting with the n vertices but no edges between them, one joins vertices together in order of the weights of vertex pairs, ignoring the edges of the original network. One can pause at any stage in this process and observe the pattern of components formed by the connections added so far, which are taken to be the communities of the network at that stage. The heirarchical clustering method thus defines not just a single decomposition of the network into communities, but a nested hierarchy of possible decompositions, having varying numbers of communities. This hierarchy can be represented as a tree or "dendrogram", an example of which is shown in Fig. 4.2. A horizontal cut through the dendrogram at any given height, such as that denoted by the dotted line in Fig. 4.2, splits the tree into the communities for the corresponding stage in the hierarchical clustering process. By varying the height of the cut, one can arrange for the number communities to take any desired value.

The construction of dendrograms is a popular technique for the analysis of network data, particularly within the sociological community. Software packa-

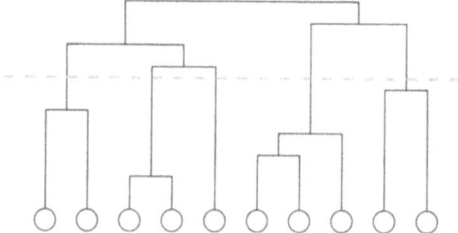

Fig. 4.2. An example of a dendrogram showing the hierarchical clustering of ten vertices. A horizontal cut through the dendrogram, such as that denoted by the dotted line, splits the vertices into a set of communities, five in this case

ges for network analysis, such as Pajek and UCInet, incorporate hierarchical clustering as a standard feature: for any network one can calculate a huge variety of vertex–vertex weights of different types and construct the corresponding dendrogram for any of them. The method however has some problems. There are many cases in which networks have rather obvious community structure, but hierarchical clustering fails to find it. One particular pathology that is frequently observed is that peripheral vertices tend to get disconnected from the bulk of the network, rather than being associated with the groups or communities that they are primarily attached to. For example, if a vertex is connected to the rest of the network by only a single edge, then presumably, were one to assign it to a community, it would be assigned to the community that the single edge leads to. In many cases, however, the hierarchical clustering method will declare the vertex instead to be a single-vertex community in its own right, in complete disagreement with our intuitive ideas of community structure.

In a recent paper therefore [17] we have proposed an alternative method for detecting community structure, based on calculations of so-called edge betweenness for vertex pairs. As we will see, this method detects the known community structure in a number of networks with remarkable accuracy.

4.2.1 Edge Betweenness and Community Detection

Freeman [16] proposed a measure of centrality for the actors in a social network which he called "betweenness". The betweenness of an actor is defined to be the number of shortest paths between pairs of vertices that pass through that actor. In cases where the number p of shortest paths between a vertex pair is greater than one, each path is given an equal weight of $1/p$. Trivial algorithms for calculating betweenness take $O(mn^2)$ time to calculate betweenness for all vertices, or $O(n^3)$ time on a sparse graph (i.e., one in which the number of edges per vertex is constant in the limit of large graph size). This makes the calculation prohibitively costly on large networks. Recently however, two new algorithms have been proposed [33, 9] that both allow the same calculation to be performed faster, in time $O(mn)$, or $O(n^2)$ on a sparse graph, by eliminating needless recalculations of geodesic paths. The betweenness of a vertex gives an indication, as the name implies, of how much the vertex is "between" other vertices. If, for example, information (or anything else) spreads through a network primarily by following shortest paths, then betweenness scores will indicate through which vertices most information will flow on average. The vertices with highest betweenness are also those whose removal will result in an increase to the geodesic distance between the largest number of other vertex pairs.

Here we consider an extension of Freeman's betweenness to the edges in a network. The betweenness of an edge is defined to be the number of shortest paths between pairs of vertices that run along that edge, with paths again being given weights $1/p$ when there are $p > 1$ between a given pair of vertices. In fact, the concept of edge betweenness actually appears to predate Freeman's work on vertex betweenness, having appeared in an obscure technical report by an Amsterdam mathematician some years earlier [4]. Edge betweenness has received very little attention in other literature until recently, but it provides us with an excellent measure of which edges in a network lie between different communities. In a network with strong community structure – groups of vertices with only a few inter-group edges joining them – at least some of the inter-group edges will necessarily receive high edge betweenness scores, since they must carry the geodesic paths between vertex pairs that lie in different communities. This implies that eliminating edges with high edge betweenness from a graph will remove the inter-group edges, and hence split the graph efficiently into its different groups. This is the principle behind our method for the detection of community structure. Our algorithm is as follows.

1. We calculate the edge betweenness of every edge in the network.
2. We remove the edge with the highest betweenness score, or randomly choose one such if more than one edge ties for the honour.
3. We recalculate betweenness scores on the resulting network and repeat from step 2 until no edges remain.

The recalculation in step 3 is crucial to the method's success. When there is more than one inter-group edge between two groups of vertices, there is no guarantee that both will receive high betweenness scores; in some cases most geodesic paths with flow along one edge and only that one will receive a high score. Recalculation ensures that at some stage in the working of the algorithm each inter-group edge receives a high score and thus gets removed.

The calculation of all edge betweennesses takes time $O(mn)$, and its repetition for all m edges thus gives the algorithm a worst-case running time of $O(m^2 n)$, or $O(n^3)$ on a sparse graph. The results of the algorithm can be represented as a dendrogram, just as in traditional hierarchical clustering, although one should be aware that the construction of the tree is not logically the same: the recalculation of the betweennesses after each edge removal means that there is no single function that can be defined for each edge in the initial graph such that the resulting dendrogram is the representation of a hierarchical clustering construction carried out using that function.

4.2.2 Examples

Here we give three examples of the application of our community structure finding algorithm to different networks. The first example is a set of computer generated graphs, specifically created to test the algorithm. We created a large number of graphs of 128 vertices each, divided into four groups of 32. Edges were placed at random between vertices within the same group with probability p_{in} and between vertices in different groups with probability p_{out}, with the values of p_{in} and p_{out} chosen to make the average degree of a vertex equal to 16, and $p_{out} \leq p_{in}$. These graphs were then fed into our community structure algorithm, and we measured what fraction of the vertices were correctly classified into their communities as a function of the ratio of p_{in} to p_{out}, or equivalently the mean number z_{out} of edges from a vertex to vertices in other communities. The results are shown in Fig. 4.3. As the figure shows, the algorithm performs almost perfectly for values of z_{out} up to about 6. Beyond this point, as z_{out} approaches the value of 8 at which each vertex has as many inter-group edges as intra-group ones, the fraction of successfully classified vertices falls off sharply.

On the same plot we also show the performance of a standard hierarchical clustering algorithm based on edge-independent path counts (maxflow) on the same set of random graphs. As the figure shows, the traditional method is far inferior to our new algorithm in finding the known community structure.

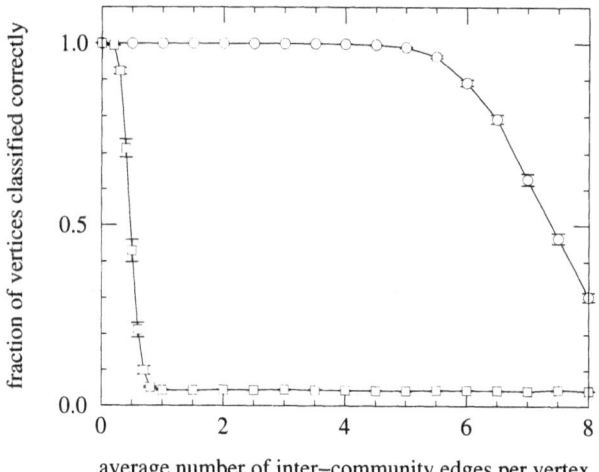

Fig. 4.3. The fraction of vertices correctly classified in applications of community structure finding algorithms to the computer-generated graphs described in the text. The circles are results for the method presented in this paper and the squares are for the standard hierarchical clustering method, using a maximum-flow measure of connection strength between vertex pairs. Each point is an average over 100 realizations of the graphs

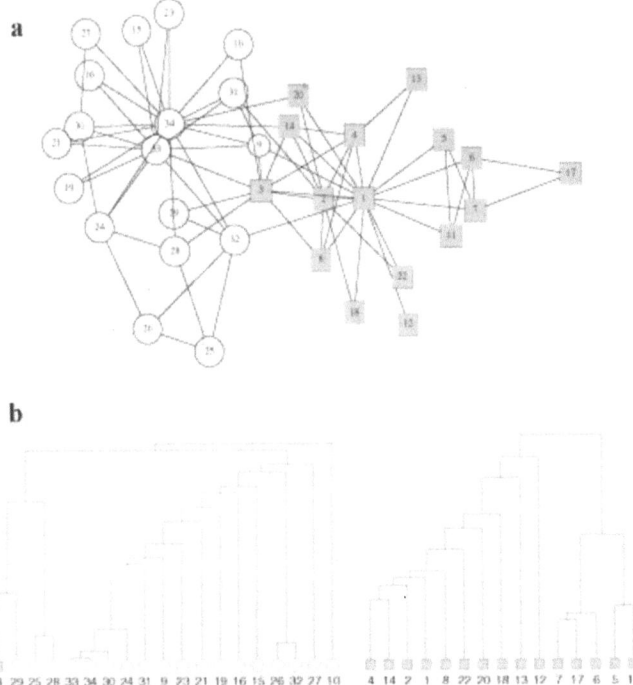

Fig. 4.4. a The friendship network given by Zachary [49] for his karate club study. Grey squares represent individuals who in the fission of the club sided with the club's instructor, while open circles represent individuals who sided with the club's president. **b** The dendrogram representing the community divisions found by our method for this network, with the same colouring scheme for the vertices

For our second example, we move to real-world network data. In 1977, Wayne Zachary published the results of an ethnographic study he had conducted of social interactions between 34 members of a karate club at an American university [49]. He recorded social contacts between members of the club over a two year period and published his results in the form of social networks. Fortuitously there arose, during the course of the study, a dispute between the two leaders of the club, the karate teacher and the club's president, over whether to raise the club's fees. Ultimately, the dispute resulted in the departure of the karate teacher and his starting another club of his own, taking with him about a half of the original club's members. Here we analyse a network constructed by Zachary of friendships between club members before the split occurred. We compare the predictions of our community-finding algorithm applied to this network with the known lines along which the club divided. Our results are shown in Fig. 4.4.

In panel (a) of the figure we show the original network, with the grey squares representing the faction that ultimately sided with the teacher (who is vertex number 1), and the open circles the faction that sided with the club's president (vertex number 34). In panel (b) we show the dendrogram output by our algorithm for this network. As the figure shows, the algorithm again performs nearly perfectly, with only one vertex, vertex number 3, being misclassified. (Inspection of panel (a) reveals that vertex 3 is in fact precisely caught in the middle of the network between the two factions, and so it is not entirely surprising that this vertex was misclassified.) Bear in mind that the network in this example was recorded *before* the fission of the club, so that the results of panel (b) are in some sense a prediction of events that were, at that time, yet to occur.

Finally, for our third example, we take a network for which we do not have any strong presuppositions about a "correct" division into communities. This example is a true experiment to see what information the algorithm can give us about a network whose structure is not wholly understood. The network in question is a food web, the web of trophic interactions (who eats whom) of marine organisms living in the Chesapeake Bay. The network was assembled by Baird and Ulanowicz [5] and contains 33 vertices representing the ecosystem's most prominent taxa. The edges in a food web are, technically, directed; they can be thought of as pointing from prey to their predators, thus indicating the direction of energy (or carbon) flow up the food chain. Here however we have ignored the directed nature of the network, considering the edges merely to be undirected indicators of trophic interaction between taxon pairs.

The dendrogram produced for this food web by our community structure algorithm is shown in Fig. 4.5. As we can see, the algorithm splits the network into two principle communities and a couple of smaller peripheral ones. We have coloured the vertices in the dendrogram to show which taxa are surface dwellers in the bay (pelagic species) and which bottom dwellers (benthic species). A few species are of undetermined status. It is clear that our algorithm has in this case primarily extracted from the network the distinction between pelagic and benthic taxa. Thus our results appear to imply that the food web in question can be split roughly into separate surface- and bottom-dwelling subsystems, with relatively weak interaction between the two. A small number of benthic species are found to belong more strongly to the pelagic community than to the benthic one, perhaps indicating that a simple classification of species by where they live is not telling the whole story for this system. The results of our analysis might also be helpful in assigning a type to the undetermined species in the network.

4.3 Origins of Community Structure and Assortative Mixing

There is certainly more than one possible explanation for the presence of community structure in a network, and different explanations may be appropriate to different networks. In the case of a social network, for example, Jin *et al.* [24] have shown that communities can arise as a result of growth dynamics of a network. If an acquaintance network grows by the introduction of pairs of people

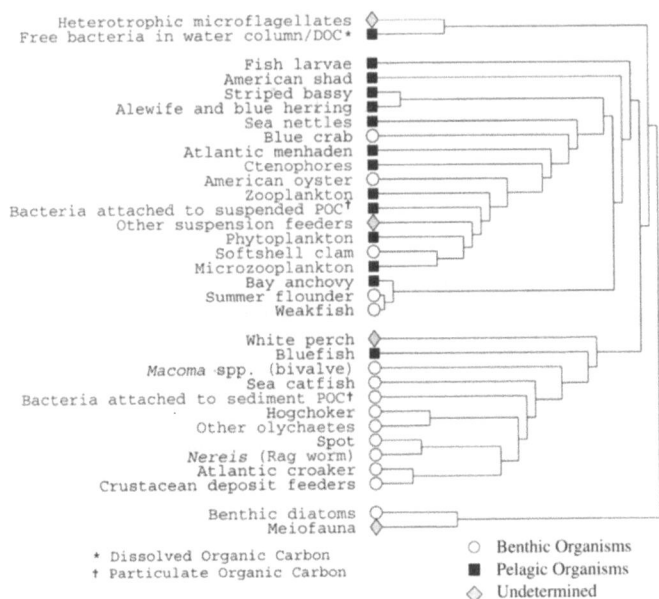

Fig. 4.5. The dendrogram found by our method for Baird and Ulanowicz's food web of marine organisms in the Chesapeake Bay [5]

to one another by a mutual acquaintance, then an initial chance acquaintance with a member of a certain community will lead to introductions to other members of that community, so that one ultimately becomes linked to many of the community's members and so becomes a member oneself. Using a simple computer model of this process, Jin *et al.* found that even networks with no initial community structure quickly develop such structure over time. One can think of this as a mechanism for the development of cliquishness in social networks.

This mechanism however is quite specific to social networks and could not be easily applied, for example, to the food web studied in the last section. It also completely ignores any personal attributes of the actors involved or affinities between actor pairs. A more general and perhaps more convincing explanation for community formation, which takes these things into account, is that of assortative mixing,[4] which is the tendency for nodes in a network to form connections preferentially to others that are like them in some way.

An example of assortative mixing in social networks is mixing by race. Table 4.1 shows data from the AMEN (AIDS in Multiethnic Neighborhoods) study [11], on mixing by race among sexual partners in the city of San Francisco, California. This part of the study focused on heterosexual partnerships,

---

[4] The name "assortative mixing" comes from the epidemiology community, where this effect has been studied extensively. It is also sometimes called "assortative matching", particularly by ecologists.

Table 4.1. The mixing matrix e_{ij} and the values of a_i and b_i for sexual partnerships in the San Francisco study described in the text. After Morris [32]

		women				a_i
		black	hispanic	white	other	
	black	0.258	0.016	0.035	0.013	0.323
men	hispanic	0.012	0.157	0.058	0.019	0.247
	white	0.013	0.023	0.306	0.035	0.377
	other	0.005	0.007	0.024	0.016	0.053
	b_i	0.289	0.204	0.423	0.084	

and the rows and columns of the matrix represent men and women in such partnerships, grouped by their (self-identified) race. Diagonal elements of the matrix represent the fraction of survey respondents in partnerships with members of their own group, and off-diagonal those in partnerships with members of other groups. Inspection of the figures shows that the matrix has considerably more weight along its diagonal than off it, indicating that assortative mixing does take place in this network. One might well expect mixing of this type to result in divisions within the community along lines of race, and we will show shortly that, within the context of simulations of network formation, assortative mixing can indeed give rise to such community structure.

The amount of assortative mixing in a network can be characterized by measuring how much of the weight in the mixing matrix falls on the diagonal, and how much off it. Let us define e_{ij} to be the fraction of all edges in a network that join a vertex of type i to a vertex of type j. In the case of the matrix of Table 4.1, where the ends of an edge always attach to one man and one woman, we should also specify which index corresponds to which type of end, which makes e_{ij} asymmetric. For example, we could specify that the first index i represents the man and the second j the woman. For networks in which there is no corresponding distinction, e_{ij} will be symmetric. The matrix should also satisfy the sum rules

$$\sum_{ij} e_{ij} = 1, \qquad \sum_j e_{ij} = a_i, \qquad \sum_i e_{ij} = b_j, \qquad (4.1)$$

where a_i and b_i are the fraction of each type of end of an edge that is attached to vertices of type i. The values of a_i and b_i for the San Francisco study are also shown in Table 4.1. On graphs where there is no distinction between the ends of edges, we will have $a_i = b_i$.

Now we can define a quantitative measure r of the level of assortative mixing in the network thus [36]:

$$r = \frac{\sum_i e_{ii} - \sum_i a_i b_i}{1 - \sum_i a_i b_i} = \frac{\mathrm{Tr}\, \mathbf{e} - \|\mathbf{e}^2\|}{1 - \|\mathbf{e}^2\|}, \qquad (4.2)$$

where $\mathbf{e}$ is the matrix whose elements are the e_{ij}, and the notation $\|\mathbf{x}\|$ indicates the sum of the elements of the matrix $\mathbf{x}$. We call the quantity r the "assortativity coefficient". It takes the value 1 in a perfectly assortative network, since in that case the entire weight of the matrix $\mathbf{e}$ lies along its diagonal and $\sum_i e_{ii} = 1$. Conversely, if there is no assortative mixing at all, then $e_{ij} = a_i b_j$ for all i, j and $r = 0$. Networks can also be disassortative: vertices may associate preferentially with others of different types – the "opposites attract" phenomenon. In that case, r will take a negative value.

One can certainly imagine that assortative mixing might apply in other types of networks as well. For example, we saw in Sect. 4.2.2 that a food web of marine organisms apparently divided into communities along lines of location – which species were surface dwellers (pelagic) and which bottom dwellers (benthic). It seems reasonable to hypothesize that the evolution of new predatory relationships between species is biased by the location of those species' living quarters, and hence that the network structure would indeed reflect the pelagic/benthic division as a result of assortative mixing by location.

We can test our hypothesis that assortative mixing could be responsible for community formation in networks by computer simulation. Given a mixing matrix of the type shown in Table 4.1, we can create a random network with the corresponding mixing pattern and any desired degree distribution by the following algorithm.

1. First we choose degree distributions $p_k^{(i)}$ for each vertex type i. The quantity $p_k^{(i)}$ here denotes the probability that a randomly chosen vertex of type i will have degree k. We can also calculate the mean degree $z_i = \sum_k k p_k^{(i)}$ for each vertex type.
2. Next we choose a size for our graph in terms of the number m of edges and draw m edges from the desired distribution e_{ij}. We count the number of ends of edges of each type i, to give the sums m_i of the degrees of vertices in each class, and we calculate the expected number n_i of vertices of each type from $n_i = m_i/z_i$ (rounded to the nearest integer).
3. We draw n_i vertices from the desired degree distribution $p_k^{(i)}$ for type i. Normally the degrees of these vertices will not sum exactly to m_i as we want them to, in which case we choose one vertex at random, discard it, and draw another from the distribution $p_k^{(i)}$, repeating until the sum does equal m_i.
4. We pair up the m_i ends of edges of type i at random with the vertices we have generated, so that each vertex has the number of attached edges corresponding to its chosen degree.
5. We repeat from step 3 for each vertex type.

We have used this algorithm to generate example networks with desired levels of assortative mixing. For example, Fig. 4.6 shows an undirected network of $n = 100$ vertices of four different types, generated using the symmetric mixing matrix

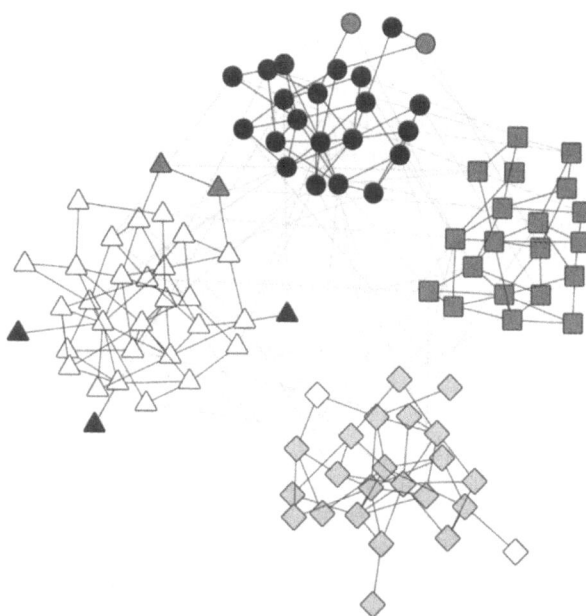

Fig. 4.6. A network generated using the mixing matrix of (4.3) and a Poisson degree distribution with mean $z = 5$. The four different shades of vertices represent the four types, and the four shapes represent the communities discovered by the community-finding algorithm of Sect. 4.2.1. The placement of the vertices has also been chosen to accentuate the communites and show where the algorithm fails. As we can see, the correspondence between vertex type and the detected community structure is very close; only nine of the 100 vertices are misclassified

$$\mathbf{e} = \begin{pmatrix} 0.18 \; 0.02 \; 0.01 \; 0.03 \\ 0.02 \; 0.20 \; 0.03 \; 0.02 \\ 0.01 \; 0.03 \; 0.16 \; 0.01 \\ 0.03 \; 0.02 \; 0.01 \; 0.22 \end{pmatrix}, \tag{4.3}$$

which gives a value of $r = 0.68$ for the assortativity coefficient. A simple Poisson degree distribution with mean $z = 5$ was used for all vertex types. The graph was then fed into the community finding algorithm of Sect. 4.2.1 and a cut through the resulting dendrogram performed at the four-community level. The communities found are shown by the four shapes of vertices in the figure and correspond very closely to the real vertex type designations, which are represented by the four different vertex shades. In other words, by introducing assortative mixing by vertex type into this network, we have created vertex-type communities that register in our community finding algorithm in exactly the same way as communities in naturally occurring networks. This strongly suggests that assortative mixing could indeed be an explanation for the occurrence of such communities, although it is worth repeating once again that other explanations are also possible.

4.4 Other Types of Assortative Mixing

Assortative mixing can depend on vertex properties other than the simple enumerative properties discussed in the preceding section. For example, we can also
have assortative mixing by scalar characteristics, either discrete or continuous.
A classic example of such mixing, much studied in the sociological literature,
is acquaintance matching by age. In many contexts, people appear to prefer to
associate with others of approximately the same age as themselves. As an example of such mixing, consider Fig. 4.7, which shows the ages at marriage of the
male and female members of 1141 married couples drawn from the US National
Survey of Family Growth [14]. Each point in the figure represents one couple, its
position along the horizontal and vertical axes corresponding to the ages of the
husband and wife respectively. The study was based on interviews with women,
and was limited to those of childbearing age, so the vertical axis cuts off aro-
und 40. Also only the first marriage for each woman interviewed is shown, even
if she married more than once. Despite these biases however, the figure reveals
a clear trend: people prefer to marry others of an age close to their own.

It is perhaps stretching a point a little to consider first marriage ties between
couples as forming a social network, since people have at most one first marriage
and hence would have a maximum degree of one within the network. Here,
however, we consider marriage age as a proxy for the ages of sexual partners in
general, and conjecture that a similar age preference will be seen in non-married
partners also, although we are not aware of any specific data to that effect.

Assortative mixing according to scalar characteristics can result in the formation of communities, just as in the case of discrete characteristics. One could have
separate communities formed of old and young people, for instance. However, it

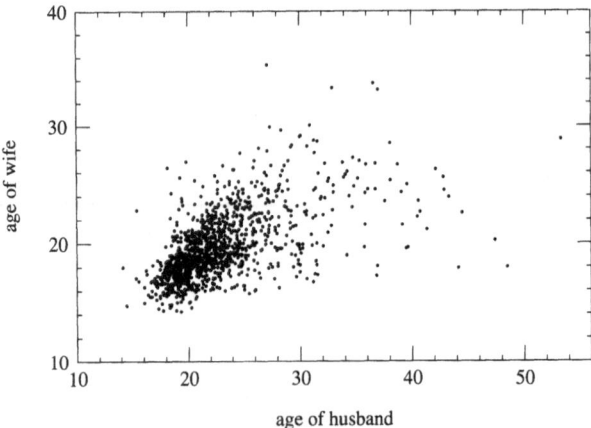

Fig. 4.7. Scatter plot of the ages at first marriage of 1141 women interviewed in the
1995 National Survey of Family Growth, and their spouses. Only women of up to 45
years of age were interview, so the vertical axis does not extend as far as the horizontal
one

is also possible that we do not get well-defined communities, but instead get an overlapping set of groups with no clear boundaries, ranging for example from low age to high age. In the sociological literature such a continuous gradation of one community into another is called "stratification" of the network.

As with assortative mixing on discrete characteristics, one can define an assortativity coefficient to quantify the extent to which mixing is biased according to scalar vertex properties. To do this, we define e_{xy} to be the fraction of edges in our network that connect a vertex of property x (e.g., age) to another of property y. The matrix e_{xy} must satisfy sum rules as before, of the form

$$\sum_{xy} e_{xy} = 1, \qquad \sum_{y} e_{xy} = a_x, \qquad \sum_{x} e_{xy} = b_y, \qquad (4.4)$$

where a_x and b_y are, respectively, the fraction of edges that start and end at vertices with ages x and y. Then the appropriate definition for the assortativity coefficient is

$$r = \frac{\sum_{xy} xy(e_{xy} - a_x b_y)}{\sigma_a \sigma_b}, \qquad (4.5)$$

where σ_a and σ_b are the standard deviations of the distributions a_x and b_y. The reader will no doubt recognize this definition of r as the standard Pearson correlation coefficient for the quantities x and y. It takes values in the range $-1 \leq r \leq 1$ with $r = 1$ indicating perfect assortative mixing, $r = 0$ indicating no correlation between x and y, and $r = -1$ indicating perfect disassortative mixing, i.e., perfect anticorrelation between x and y.

If we take the marriage data from Fig. 4.7, for example, and feed it into (4.5), we find that $r = 0.57$, indicating once again that mixing is strongly assortative (as is in any case obvious from the figure).

Mixing could also depend on vector or even tensor characteristics of vertices. One example would be mixing by geographical location, which could be regarded as a two-vector. It seems highly likely that if one were to record both acquaintance patterns and geographical location for actors in a social network, one would discover that acquaintance is strongly dependent on geography, with people being more likely to know others who live in the same part of the world as themselves.

4.4.1 Mixing by Vertex Degree

We will spend the rest of this article examining one particular case of mixing according to a scalar vertex property, that of mixing by vertex degree, which has been studied for some time in the social networks literature and has recently attracted attention in the mathematical and physical literature also. Krapivsky and Redner [27] for instance found in studies of the preferential attachment model of Barabási and Albert [6] that edges did not fall between vertices independent of their degrees. Instead there was a higher probability to find some degree combinations at the ends of edges than others. Pastor-Satorras *et al.* [40] subsequently showed for data on the structure of the Internet at the level of autonomous systems that the degrees of adjacent vertices were anticorrelated, i.e., that high-degree vertices prefer to attach to low-degree vertices, rather than other high-degree ones – the network is disassortative by degree. To demonstrate this, they measured the mean degree degree $\langle k_{nn} \rangle$ of the nearest-neighbours of a vertex, as a function of that vertex's degree k. They found that $\langle k_{nn} \rangle$ *decreases* with increasing k, approximately as $k^{-1/2}$. That is, the mean degree of your neighbours goes down as yours goes up. Maslov and Sneppen [29] have offered an explanation of this result in terms of ensembles of graphs in which double edges between vertices are forbidden. Maslov and Sneppen also showed in a separate paper [30] that the protein interaction network of the yeast *S. Cerevisiae* displays a similar sort of disassortative mixing.

An alternative way to quantify assortative mixing by degree in a network is to use an assortativity coefficient of the type described in the previous section [35]. Let us define e_{jk} to be the fraction of edges in a network that connect a vertex of degree j to a vertex of degree k. (As before, if the ends of an edge connect different types of vertices, then the matrix will be asymmetric, otherwise it will be symmetric.) In fact, we define j and k to be the "excess degrees" of the two vertices, i.e., the number of edges incident on them less the one edge that we are looking at at present. In other words, j and k are one less than the total degrees of the two vertices. This designation turns out to be mathematically convenient for many developments. If the degree distribution of the network as a whole is p_k, then the distribution of the excess degree of the vertex at the end of a randomly chosen edge is

$$q_k = \frac{(k+1)p_{k+1}}{z}, \tag{4.6}$$

where $z = \sum_k k p_k$ is the mean degree [37]. Then one can define the assortativity coefficient to be

$$r = \frac{\sum_{jk} jk(e_{jk} - q_j q_k)}{\sigma_q^2}, \tag{4.7}$$

where σ_q is the standard deviation of the distribution q_k. On a directed or similar network, where the ends of an edge are not the same and e_{jk} is asymmetric, this generalizes to

$$r = \frac{\sum_{jk} jk(e_{jk} - q_j^a q_k^b)}{\sigma_a \sigma_b}, \tag{4.8}$$

where σ_a and σ_b are the standard deviations of the distributions q_k^a and q_k^b for the two types of ends. (The measure introduced by Pastor-Satorras *et al.* [40] can also be expressed simply in terms of the matrix e_{jk}: it is $\langle k_{nn} \rangle = \sum_j j e_{jk}$. Maslov and Sneppen [30, 29] gave entire plots of the raw e_{jk}, using colours to

Table 4.2. Size n and degree assortativity coefficient r for a number real-world networks. Social networks: coauthorship networks of (a) physicists and biologists [34] and (b) mathematicians [19]; (c) collaborations (co-starring relationships) of film actors [46, 39]; (d) directors of Fortune 1000 companies for 1999, in which two directors are connected if they sit on the board of directors of the same company [13, 39]; (e) network of email address books of computer users [38]. Technological networks: (f) network of direct peering relationships between autonomous systems on the Internet, April 2001 [12]; (g) network of hyperlinks between pages in the World-Wide Web domain nd.edu *circa* 1999 [3]; (h) network of dependencies between software packages in the GNU/Linux operating system [36]. Biological networks: (i) protein–protein interaction network in the yeast *S. Cerevisiae* [22]; (j) metabolic network of the bacterium *E. Coli* [23]; (k) neural network of the nematode worm *C. Elegans* [47, 46]; tropic interactions between species in the food webs of (l) Ythan Estuary, Scotland [21] and (m) Little Rock Lake, Wisconsin [28]. After Newman [36]

	network	type	size n	assortativity r	ref.
social	physics coauthorship	undirected	52 909	0.363	a
	biology coauthorship	undirected	1 520 251	0.127	a
	mathematics coauthorship	undirected	253 339	0.120	b
	film actor collaborations	undirected	449 913	0.208	c
	company directors	undirected	7 673	0.276	d
	email address books	directed	16 881	0.092	e
technol.	Internet	undirected	10 697	−0.189	f
	World-Wide Web	directed	269 504	−0.067	g
	software dependencies	directed	3 162	−0.016	h
biological	protein interactions	undirected	2 115	−0.156	i
	metabolic network	undirected	765	−0.240	j
	neural network	directed	307	−0.226	k
	marine food web	directed	134	−0.263	l
	freshwater food web	directed	92	−0.326	m

code for different values. These plots are however rather difficult to interpret by eye.)

In Table 4.2 we show values of r measured for a variety of different real-world networks. The networks shown are divided into social, technological and biological networks, and a particularly striking feature of the table is that the values of r for the social networks are all positive, indicating assortative mixing by degree, while those for the technological and biological networks are all negative, indicating disassortative mixing. It is not clear at present why this should be, although explanations for the observed mixing behaviours have been proposed in some specific cases [29, 36].

As with the mixing by discrete enumerative characteristics discussed in Sect. 4.3, we can also investigate the effects of assortative mixing by looking at computer generated networks with particular types of mixing. Unfortunately, no simple algorithm exists for generating graphs mixed by vertex degree analogous to that of Sect. 4.3 (see Dorogovtsev *et al.* in this volume and Newman [36]) and one is forced to resort to Monte Carlo generation of graphs using

(a)   (b)

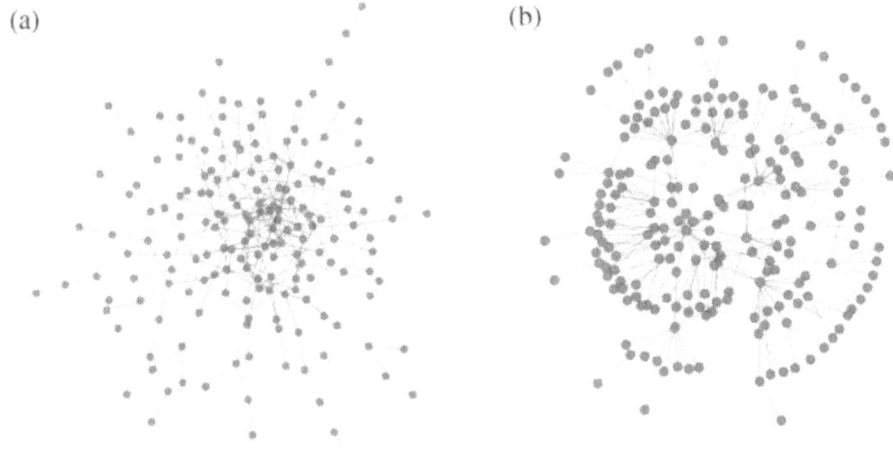

Fig. 4.8. The giant component of two graphs generated using a Monte Carlo procedure with edge distribution given by (4.9) with $\kappa = 10$ and **a** $p = 0.5$ and **b** $p = 0.05$

Metropolis–Hasting type algorithms of the sort widely used for graph generation in mathematics and quantitative sociology. Such algorithms however are straightforward to implement. For the present case, we take the simple example form

$$e_{jk} = \mathcal{N} \mathrm{e}^{-(j+k)/\kappa} \left[\binom{j+k}{j} p^j q^k + \binom{j+k}{k} p^k q^j \right], \qquad (4.9)$$

where $p + q = 1$, $\kappa > 0$, and $\mathcal{N} = \frac{1}{2}(1 - \mathrm{e}^{-1/\kappa})$ is a normalizing constant. This means that the distribution of the sum $j + k$ of the excess degrees at the ends of an edge falls off as a simple exponential, while that sum is distributed between the two ends binomially, the parameter p controlling the assortative mixing. For values of p ranging from 0 to $\frac{1}{2}$ we get various values of the assortativity r, both positive and negative, passing through zero at $p_0 = \frac{1}{2} - \frac{1}{4}\sqrt{2} = 0.1464\ldots$

As an example, we show in Fig. 4.8 the giant components of two graphs of this type generated using the Monte Carlo method. One of them, graph (a), is assortatively mixed by degree, while the other, graph (b), is disassortatively mixed. The difference between the two is clear to the eye. In the first case, because the high degree vertices prefer to attach to one another, there is a central "core" to the network, composed of these high-degree vertices, and a straggling periphery of low-degree vertices around it. In epidemiology a dense central portion of this type is called a "core group" and is thought to be capable of acting as a reservoir for disease, keeping diseases circulating even when the density of the network as a whole is too low to maintain endemic infection. In social network analysis one also talks of "core/periphery" distinctions in networks, another concept that mirrors what we see here. In the second graph, which is disassortative, a contrasting picture is evident: the high-degree vertices prefer not to associate with

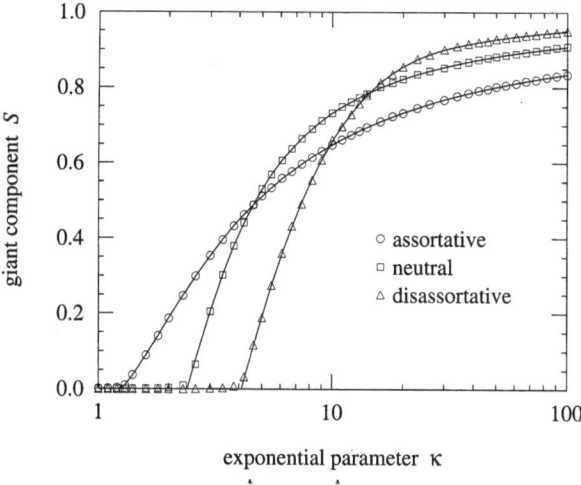

Fig. 4.9. The size of the giant component as a function of graph size for graphs with the edge distribution given in (4.9), for three different values of the parameter p, which controls the assortativity. The points are simulation results for graphs of $N = 100\,000$ vertices while the solid lines are the analytic solution for the same quantity given by Newman [35]. Each point is an average over ten graphs; the resulting statistical errors are smaller than the symbols. The values of p are 0.5 (*circles*), $p_0 = 0.146\ldots$ (*squares*), and 0.05 (*triangles*)

one another, and are as a result scattered widely over the network, producing a more uniform appearance.

To shed more light on the effects of assortativity, we show in Fig. 4.9 the size of the largest component in networks of this type as the degree distribution parameter κ is varied, for various values of p. For low values of κ the mean degree of the network is small, and the resulting density of edges is too low to produce percolation in the network, so there is no giant component. As κ increases, however, there comes a point, clearly visible on the plot, at which the edge density is great enough to form a giant component. Figure 4.9 reveals two interesting features of this transition. First, the position of the transition, the value of the parameter κ at which it takes place, is smaller in assortatively mixed networks than in disassortative ones. In other words, it appears that the presence of assortativity in the degree correlation pattern allows the network to percolate more easily. This result is intuitively reasonable: the core group of the assortative network seen in Fig. 4.8a has a higher density of edges than the network as a whole and so one would expect percolation to take place in this region before it would in a network with the same average density but no core group.

Second, the figure shows that, even though the assortative network percolates more easily than its disassortative counterpart, its largest component does not grow as large as that of the disassortative network in the limit where κ becomes large. This too can be understood in simple terms: percolation occurs more easily when there is a core group, but is also largely confined to that core group and so does not spread to as large a portion of the network as it would in other cases.

In epidemiological terms, one could think of these two results as indicating that assortative networks will support the spread and persistence of a disease more easily than disassortative ones, because they possess a core group of connected high-degree vertices. But the disease is also restricted mostly to that core group. In a disassortative network, although percolation and hence epidemic disease requires a denser network to begin with, when it does happen it will affect a larger fraction of the network, because it is not restricted to a core group.

4.5 Conclusions

In this article we have examined two related properties of networks: community structure and assortative mixing. We have described a new algorithm for finding groups of tightly-knit vertices within networks – communities in our nomenclature – which is based on the calculation of an "edge betweenness" index for network edges. The algorithm appears to be successful at detecting known community structure in various example networks, and we have found that many real-world networks do indeed possess community structure to a greater or lesser degree.

Turning to possible explanations for this structure we have suggested that assortative mixing, the preferential association of vertices in a network with others that are like them in some way, is one possible mechanism for community formation. We have defined a measure of the strength of assortative mixing and applied it, for example, to data on mixing by race in social networks, showing that there is strong assortativity in this case, at least for the survey data that we have examined. We have also given a simple algorithm for creating networks with assortative mixing according to discrete characteristics imposed upon the vertices, and used it to generate example networks which, when fed into our community detection algorithm, reveal strong community structure similar to that seen in the real-world data. This lends some conviction to the theory that assortative mixing could, at least in some cases, be a contributing factor in the formation of communities within networks.

We have also looked at assortative mixing by scalar characteristics of vertices, such as the age of individuals in a social network, and particularly vertex degree. By measuring mixing of the latter type for a variety of different networks, we have shown that social networks appear often to be assortatively mixed by degree, while technological and biological networks appear normally to be disassortative. Using computer generated model networks we have also shown that assortativity by vertex degree makes networks percolate more easily – they develop a giant component for a lower average edge density than a similar network with neutral or disassortative mixing. Conversely, however, disassortative networks tend to have larger giant components when they do develop. These findings have implications for epidemiology, for example: they imply that a disease spreading on a network that is assortatively mixed, as most social networks appear to be, would reach epidemic proportions more easily than on a disassortative network, but that an epidemic might ultimately affect fewer people than in the disassortative case.

Looking ahead, some obvious next steps in the studies presented here are the application of community finding algorithms to other networks, the study of mixing patterns in other networks, and theoretical investigations of the effects of assortative mixing and other network correlations on network structure and function, including for instance network resilience and network epidemiology. A number of authors have already started work on these problems [20, 48, 8, 43, 35, 36].

Acknowledgements

The authors would like to thank Jennifer Dunne, Neo Martinez and Doug White for help assembling and interpreting the data used in Figs. 4.4 and 4.5, and László Barabási, Jerry Davis, Jennifer Dunne, Jerry Grossman, Hawoong Jeong, Neo Martinez and Duncan Watts for providing data used in the calculations for Table 4.2. The marriage data for Fig. 4.7 were provided by the Inter-University Consortium for Political and Social Research at the University of Michigan. This work was supported in part by the National Science Foundation under grant DMS–0109086.

References

1. L.A. Adamic, R.M. Lukose, A.R. Puniyani, B.A. Huberman: Phys. Rev. E **64**, 046135 (2001)
2. R. Albert, A.-L. Barabási: Rev. Mod. Phys. **74**, 47 (2002)
3. R. Albert, H. Jeong, A.-L. Barabási: Nature **401**, 130 (1999)
4. J.M. Anthonisse: The rush in a directed graph. Technical Report BN 9/71, Stichting Mathematicsh Centrum, Amsterdam (1971)
5. D. Baird, R.E. Ulanowicz: Ecological Monographs **59**, 329 (1989)
6. A.-L. Barabási, R. Albert: Science **286**, 509 (1999)
7. A. Barbour, D. Mollison: In: *Stochastic Processes in Epidemic Theory*, edited by J.P. Gabriel, C. Lefevre, P. Picard (Springer, New York, 1990), pp. 86–89
8. M. Boguna, R. Pastor-Satorras, A. Vespignani: Absence of epidemic threshold in scale-free networks with connectivity correlations. Preprint cond-mat/0208163 (2002)
9. U. Brandes: Journal of Mathematical Sociology **25**, 163 (2001)
10. A. Broder, R. Kumar, F. Maghoul, P. Raghavan, S. Rajagopalan, R. Stata, A. Tomkins, J. Wiener: Computer Networks **33**, 309 (2000)
11. J.A. Catania, T.J. Coates, S. Kegels, M.T. Fullilove: Am. J. Public Health **82**, 284 (1992)

12. Q. Chen, H. Chang, R. Govindan, S. Jamin, S.J. Shenker, W. Willinger: In: *Proceedings of the 21st Annual Joint Conference of the IEEE Computer and Communications Societies* (IEEE Computer Society, 2002)

13. G.F. Davis, M. Yoo, W.E. Baker: The small world of the corporate elite. Preprint, University of Michigan Business School (2001)

14. *National Survey of Family Growth, Cycle V, 1995* (U.S. Department of Health and Human Services, National Center for Health Statistics, Hyattsville, MD, 1997)

15. S.N. Dorogovtsev, J.F.F. Mendes: Advances in Physics **51**, 1079 (2002)

16. L. Freeman: Sociometry **40**, 35 (1977)

17. M. Girvan, M.E.J. Newman: Proc. Natl. Acad. Sci. USA **99**, 8271 (2002)

18. P. Grassberger: Math. Biosci. **63**, 157 (1983)

19. J.W. Grossman, P.D.F. Ion: Congressus Numerantium **108**, 129 (1995)

20. P. Holme, M. Huss, H. Jeong: Subnetwork hierarchies of biochemical pathways. Preprint cond-mat/0206292 (2002)

21. M. Huxham, S. Beany, D. Raffaelli: Oikos **76**, 284 (1996)

22. H. Jeong, S. Mason, A.-L. Barabási, Z.N. Oltvai: Nature **411**, 41 (2001)

23. H. Jeong, B. Tombor, R. Albert, Z.N. Oltvai, A.-L. Barabási: Nature **407**, 651 (2000)

24. E.M. Jin, M. Girvan, M.E.J. Newman: Phys. Rev. E **64**, 046132 (2001)

25. J.M. Kleinberg: In: *Proceedings of the 32nd Annual ACM Symposium on Theory of Computing* (Association of Computing Machinery, New York, 2000), pp. 163–170

26. J.M. Kleinberg: In: *Proceedings of the 2001 Conference on Neural Information Processing Systems*, edited by T.G. Dietterich, S. Becker, Z. Ghahramani (MIT Press, Cambridge, MA, 2002)

27. P.L. Krapivsky, S. Redner: Phys. Rev. E **63**, 066123 (2001)

28. N.D. Martinez: Ecological Monographs **61**, 367 (1991)

29. S. Maslov, K. Sneppen: Pattern detection in complex networks: Correlation profile of the Internet. Preprint cond-mat/0205379 (2002)

30. S. Maslov, K. Sneppen: Science **296**, 910 (2002)

31. Y. Moreno, R. Pastor-Satorras, A. Vespignani: Eur. Phys. J. B **26**, 521 (2002)

32. M. Morris: In: *Epidemic Models: Their Structure and Relation to Data*, edited by D. Mollison (Cambridge University Press, Cambridge, 1995), pp. 302–322

33. M.E.J. Newman: Phys. Rev. E **64**, 016132 (2001)

34. M.E.J. Newman: Proc. Natl. Acad. Sci. USA **98**, 404 (2001)

35. M.E.J. Newman: Assortative mixing in networks. Preprint cond-mat/0205405 (2002)

36. M.E.J. Newman: Mixing patterns in networks. Preprint cond-mat/0209476 (2002)

37. M.E.J. Newman: In: *Handbook of Graphs and Networks*, edited by S. Bornholdt, H.G. Schuster (Wiley-VCH, Berlin, 2002)

38. M.E.J. Newman, S. Forrest, J. Balthrop: Phys. Rev. E **66**, 035101 (2002)

39. M.E.J. Newman, S.H. Strogatz, D.J. Watts: Phys. Rev. E **64**, 026118 (2001)

40. R. Pastor-Satorras, A. Vázquez, A. Vespignani: Phys. Rev. Lett. **87**, 258701 (2001)

41. J. Scott: *Social Network Analysis: A Handbook* (Sage Publications, London, 2000), 2nd ed.

42. S.H. Strogatz: Nature **410**, 268 (2001)

43. A. Vazquez, Y. Moreno: Resilience to damage of graphs with degree correlations. Preprint cond-mat/0209182 (2002)

44. S. Wasserman, K. Faust: *Social Network Analysis* (Cambridge University Press, Cambridge, 1994)

45. D.J. Watts, P.S. Dodds, M.E.J. Newman: Science **296**, 1302 (2002)
46. D.J. Watts, S.H. Strogatz: Nature **393**, 440 (1998)
47. J.G. White, E. Southgate, J.N. Thompson, S. Brenner: Phil. Trans. R. Soc. London **314**, 1 (1986)
48. D. Wilkinson, B.A. Huberman: A method for finding communities of related genes. Preprint, Stanford University (2002)
49. W.W. Zachary: Journal of Anthropological Research **33**, 452 (1977)

5 Effect of Accelerated Growth on Networks Dynamics

J.F.F. Mendes

Departamento de Física, Universidade de Aveiro, Campus Universitário de Santiago, 3810-193 Aveiro, Portugal

Abstract. In most of real growing networks the mean number of connections per vertex increases with time. Among the examples of large networks presenting this type of growth are the Internet, the Word Wide Web, collaborations networks, and many others. We call this type of growth *accelerated growth*. We show that the accelerated growth influences the distribution of connections and as consequence it may determine the structure of a network. For the growing networks with preferential linking and increasing density of links, two scenarios are possible. In one of them, the value of the exponent γ of the connectivity distribution $P(q,t) \propto q^{-\gamma}$ is between 3/2 and 2. In the other the exponent is, $\gamma > 2$, and the distribution is necessarily non-stationary. We discuss the general consequences of the acceleration and demonstrate its features applying it to simple illustrating examples. In particular, we show that the accelerated growth fairly well explains the structure of the Word Web (the network of interacting words of human language).

5.1 The Meaning of Acceleration in Networks

In the last recent years there has been a growing interest in the study of topologic and dynamical properties of what has come to be known as complex networks. Many models have been devoted to the study of real networks. Most of models of evolving networks contain a very important assumption. In these models it is assumed that the total number of edges of a growing network is a linear function of its size (total number of vertices). The linear growth does not change the average degree of the network [1, 2, 3] This same is not true in the case of accelerated growth. We present some examples of networks that belong to the family of nets that has an acceletated growth. However, there are also examples of real nets that do not belong to this family and also not to the linear growth ones, is the case, for instance, of biological networks.

The first model for the growth of networks under mechanism of preferential linking which was introduced by Barabási-Albert (Barabási-Albert model) [4] (see also [5]), is only one example of a linearly growing network from a very long list [6, 7, 8, 9, 10, 11, 12, 13]. Thus, a linear type of growth is usually supposed to be a natural feature of growing networks. But let us ask ourselves, whether this very particular case, that is, the linear growth is or not so widespread in real networks. To answer this question we must look at existing empirical data. In any case we will consider these problems using general arguments. Assuming

As we can see from the Table 5.1, the average degree of the WWW is increa-

The Internet:

ghly speaking, the Internet (support of the WWW) is a set of vertices, linked
vires. The vertices of the Internet are hosts (computers of users), servers
nputers or programs providing a network service that also may be hosts),
routers that arrange traffic across the Internet. Connections are naturally
irected (an undirected network, the physical connection (wires) can transport
rmation in both sides). In January of 2001, the Internet contained already
ut 100 millions hosts. One should emphasize, that it is not the hosts that
rmine the structure of the Internet, but rather, routers and domains. In
· of 2000, there were about 150 000 routers in the Internet [15]. Later, the
iber rose to 228 265 (data from [17]). Thus, one can consider the topology of
Internet on a router level or inter-domain topology [18]. In the latter case
er-domain level), it is actually a small network not allowing to make a good
lysis (see Table 5.2).

The last data of [18] are for December of 1998. However, one may use more
ent data on "autonomous systems". Extensive data on connections of ope-
ng "autonomous systems" (AS) in the Internet are being collected by the
ional Laboratory for Applied Network Research (NLANR). For nearly each
, starting from November of 1997, NLANR has a map of connections of AS.
ese maps are closely related to the Internet graph on the inter-domain level.
tistical analysis of these data was made in [19, 20]. The data were averaged,
l for 1997 the average degree 3.42 was obtained; in 1998, the average degree
s 3.65; in 1999, $\overline{k} = 3.76$. Again we see that the average degree of the Internet
the inter-domain level (more rigorously speaking, on the AS level) is increa-
g. One should add that the growth of the average degree of the net of AS was
o indicated in [21]. Table 5.2 summarises some of the known values for the
ernet size on time.

It is clear that the average degree of the Internet on the inter-domain level
increasing in time.

Table 5.2. Known values of the Internet size (inter-domain level) at different times

date	N_{nodes}	N_{links}	$\overline{k}$
Nov 97	3015	5156	3.42
Apr 98	3530	6432	3.64
Dec 98	4389	8256	3.76
Dec 99	6374	13641	4.28
Sep 01	11927	27492	4.61

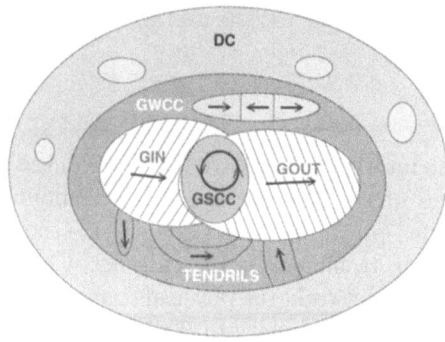

Fig. 5.1. The WWW is a directed graph. In this schematic picture w
ture when the strongly connected component (GSCC) is present. The
components, are the giant out- and in-component, respectively GOU
giant weekly connected component (GWCC)....

that the network is scale-free, we will describe the possible degre
and show , in this case, that the total number of links is a power-
the network size. Let us present results of some of the most well kn

(i) *The World Wide Web:*
The WWW consists of a set of documents (pages) plus hyper-links
The WWW contrarily to the Internet (that will be presented bel
ted network. Although hyper-links are directed, pairs of counter-li
ple, may produce undirected connections. Links inside pages (self-
usually not considered as edges of the WWW, so this network do
"tadpoles" (closed one-edge loops). Figure 5.1 shows in a schem
structure of a directed graph and all his components (for a detaile
see [3]).

According to [14], in May of 1999, and using data from Altavist
consisted of 203×10^6 vertices (URLs, i.e., pages) and $1466 \times 10^{!}$
So, the average in- and out- degree were $\bar{k}_i = \bar{k}_o = 7.22$. It is poss
values for the WWW size for other times. We summarise some kr
Table 5.1.

The average in- and out-degrees are equal to each other, since al
tions are inside the WWW. (Notice that the "physical" time is uni
us, so that, in principle, we might not mention any date.)

Table 5.1. Known values of the size of WWW at different tir

date	N_{nodes}	N_{links}	$\bar{k}$
May 99	203×10^6	1466×10^6	7.22
Oct 99	271×10^6	2130×10^6	7.85

(iii) *Collaboration Networks:*

Roughly speaking a collaboration network consists of a set of vertices and links, where vertices are collaborators and a pair of vertices is connected together by an undirected edge if there is at least one act of collaboration between them [4, 24]. For example, in scientific collaboration networks (networks of coauthorships), vertices are authors, and edges are coauthorships [25]. Another example, are all Hollywood movies; if we assign a node to each actor and connect two nodes if the corresponding actors have worked together in one or more movies, we obtain a collaboration network. Such networks are projections of more complex and informative bipartite graphs, which contain two types of vertices: collaborators and acts of collaboration. Each collaborator is connected to all the acts of collaboration, in which he was involved.

Empirical data of [26, 28] for large scientific collaboration networks indicate the linear growth of their average degree with the increasing number of their vertices. This means that the total number of edges in a network increases as a square of the total number of vertices.

Thus we see that the accelerated growth of networks is not an exception but rather a rule. On the contrary, the linear growth is a simple but very particular case.

5.2 Degree Distributions

5.2.1 What Types of Degree Distribution Can We Have?

We focus here our attention on the simplest degree distributions of networks, $P(k)$. Most of empirical results are obtained for this simple characteristic. Unfortunately, the degree distribution (in-, out-degree distribution) is a restricted characteristic of networks. Indeed, degree is a one-vertex quantity, so that, in general, degree distribution does not yield information about the global topology of a network. For a more detailed understanding one should find correlations between nodes.

In most of cases, for example, for growing networks, in which correlations between degrees of vertices are strong [11, 19, 20], a degree distribution is only the tip of the iceberg. The Internet corresponds to one such example. Basically its correlations arise because of its hierarchical structure. As a consequence, vertices with high degrees are expected to be connected to vertices with small degrees. Of course, if degree-degree correlations in a network are absent, then, knowing the degree distribution of a network, one can completely characterize the net . We face this situation in many equilibrium networks.

Furthermore, analytical results on percolation on networks [31, 32], disease spread within them [34, 35], etc. were obtained just for a simple construction without degree-degree correlations. This construction is a standard model of a maximally random graph with an arbitrary degree distribution taken from

mathematical graph theory ("random graphs with restricted degree sequences") [36]. Luckily, it seems that main percolation and disease spread results
that was obtained for equilibrium networks are still valid for non-equilibrium
nets.

So we can ask, what kinds of degree distributions can be observed in networks? Here we list the main types with some simple examples of the corresponding networks.

(a) Poisson degree distribution: $P(k) = e^{-\bar{k}}\bar{k}^k/k!$ (see Fig. 5.2a)
The Poisson distribution is realized in a classical random equilibrium graph of
Erdös and Rényi [37, 38] in the limit of the infinite network, that is, when the
total number of vertices N is infinite. Pairs of randomly chosen vertices are
connected by edges. One may create at random L edges in the graph, or connect
pairs of vertices with the probability $L/[N(N-1)/2]$. In both these cases, the
resulting graph is the same in the limit $N \to \infty$.

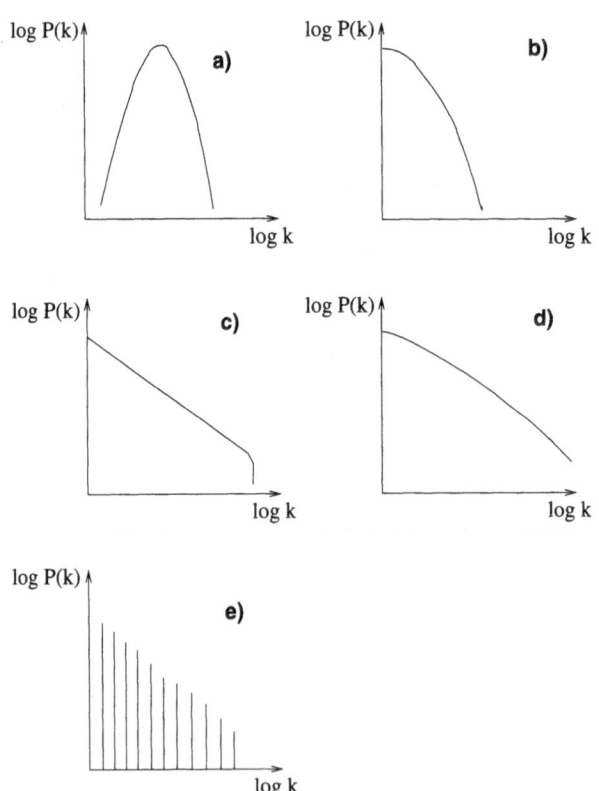

Fig. 5.2. "Zoology" of degree distributions in networks. Main types of a degree distribution in log-log plots. **a** Poisson, **b** exponential, **c** power-law, **d** multifractal, and
e discrete distributions

(b) Exponential degree distribution: $P(k) \sim \exp(-k/\text{const})$ (see Fig. 5.2b)
A citation graph[1] with attachment of new vertices to randomly chosen old ones produces the exponential distribution, but this is only one possible example. (Let each new vertex have the same number of connections, that is, the growth is linear.)

Also, the exponential degree distribution is rather usual for many equilibrium networks that are constructed by mechanism of preferential linking.

(c) Power-law degree distribution: $P(k) \sim k^{-\gamma}$ (see Fig. 5.2c)
Here the standard example is the Barabási-Albert model [4] (see also [5]). This growing network is a linearly growing citation graph in that new vertices are attached to preferentially chosen old ones. "Popular" old vertices attract more new connections than "failures": "*popularity is attractive*". This is a quite general principle. For example, this one is incorporated in the Simon model [39, 40]. In the Barabási-Albert model, the probability that an edge becomes attached to some vertex is proportional to the degree k of this vertex. This yields $\gamma = 3$. If the probability is proportional to $k + \text{const}$ (a linear preference function), γ takes values between 2 and ∞ as the constant changes from -1 to ∞ [7].

Power-law distributions are usually called scale-free or fractal.

(d) Multifractal degree distributions (see Fig. 5.2c)
This distribution has a continuum spectrum of power laws with different weights. The growth of a network may produce such a degree distribution if new vertices partially copy degrees of old ones [41]. In particular, multifractal degree distributions emerge in some models of networks of protein-protein interactions [42]. Multifractal distributions is a more general case of a fat-tailed distribution than power-law distributions. Numerous empirical data were fitted by a power-law dependence. However, there were no attempts to check the possibility that at least some of empirical degree distribution are multifractal.

(e) Discrete degree distributions (see Fig. 5.2d)
Deterministic growing graphs have a discrete spectrum of degrees. Recently, it was demonstrated that some simple rules of deterministic growth may produce discrete degree distributions with a power-law decay [43]. Moreover, deterministic graphs from [3, 44, 45] have an average shortest-path length, which is proportional to the logarithm of their size. Figure 5.3 shows a simple deterministic graph [3, 44] with the discrete degree distribution that is characterized by exponent $\gamma = 1 + \ln 3/\ln 2$.

5.2.2 The Most Interesting Case: Power-Law Degree Distribution

Power-law (that is, "scale-free") degree distributions is a prominent particular case of fat-tailed degree distributions, which are widespread in real networks

---

[1] In the citation graph nodes are papers and links are the citations to previously published papers. It is a growing graph in which new links do not emerge between pairs of old nodes.

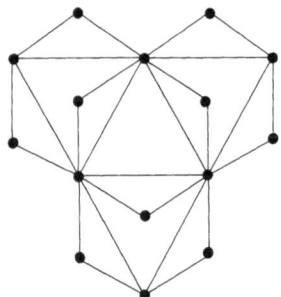

Fig. 5.3. A simple deterministic graph with a power-law discrete degree distribution. The growth starts from a single edge between two vertices. At each time step, each edge of the graph generates a new vertex, which became attached to both the end vertices of the mother edge. The average shortest-path length of this graph grows logarithmically with the total number of vertices

(both natural and artificial) [4, 5, 24]. Let us discuss briefly the general features of power-law distributions.

One may ask, what values can the exponent γ take? To answer this question one should impose the natural restriction that follows from the normalization condition $\int dk P(k) = 1$ (in this discussion we change the corresponding sum to the integral). We may not be worried about the low-degree region, since the degree distribution is certainly restricted below some characteristic degree k_0. Only the large degree behavior of the degree distribution is interesting for us. Therefore, from this condition we get that $\gamma > 1$, otherwise the integral is divergent.

If a network grows linearly, so the first moment of the distribution (the average degree $\bar{k}$), is independent of time, then we have a second restriction $\int dk k P(k) < \infty$ which implies that $\gamma > 2$ for linearly growing networks.

The finite size effect cuts the power-law part of the degree distribution at large degrees. This produces size-dependent degree distributions. One may easily estimate the position of the cutoff k_{cut} in the situation where $\gamma > 2$. Let the total number of vertices in the net be t, and k_0 be some characteristic degree, below which the distribution is, for example, constant or even zero. Then, using the normalization $\int dk P(k) = 1$ gives the power-law part of the degree distribution of the form $P(k) \sim [(\gamma - 1)k_0^{\gamma-1}]k^{-\gamma}$ for $k_0 < k < k_{cut}$.

When one measures the degree distribution of a network using only one realization of the growth process, strong fluctuations are observed at degree $k_f(t)$ that is determined by the condition $tP(k_f(t)) \sim 1$. This means that only one vertex in the network has such degree. (More rigorously speaking, the number of such vertices is of the order of one.) This is the first natural scale of the degree distribution.

One may improve the statistics by measuring many realizations of the growth process, or, for example, by passing to the cumulative distribution $P_{cum}(k) \equiv \int_k^\infty dk\, P(k)$. Both these tricks allow us to reduce the above fluc-

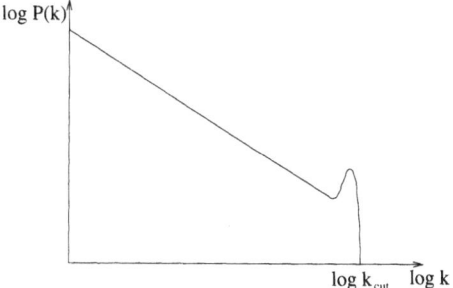

Fig. 5.4. The typical form of a power-law degree distribution of finite growing networks. The finite-size cutoff is given by (5.1). The hump near the cutoff depends on initial conditions (we do not account for the factor of mortality)

tuations. However, we still cannot surpass the next threshold that is originated from the second natural scale, k_{cut}: $tP_{cum}(k_{cut}(t)) \sim 1$. This means that only one vertex in the network is of degree greater than k_{cut}. (Again, more rigorously, the number of such vertices is of the order of one.) Using the above expression for $P(k)$ gives

$$k_{cut} \sim k_0 \, t^{1/(\gamma-1)} \, . \tag{5.1}$$

Notice that the only reason for this estimate for the cutoff is the natural scale of the problem. Hence more convincing arguments are necessary. The estimate was checked for some specific models. A growing network [13] was solved exactly, and the exact position of the cutoff have coincided with (5.1). The degree distribution of this network has a typical form (see Fig. 5.4). Notice a hump near k_{cut} in Fig. 5.4. This is a trace of initial conditions. Simulation of a scale-free equilibrium network [46] also yielded the cutoff at this point. However, the introduction of the death of vertices in the network may change the estimate (5.1). This factor also removes the hump from the degree distribution. Here we do not consider such situations.

The cutoff (5.1) hinders measurements of power-law dependences in networks [13]. From (5.1) one sees that the measurements of large enough γ are actually impossible. Indeed, in this case k_{cut} is small even for very large networks, and there is no room $\ln k_0 < \ln k < \ln k_{cut}$ for fitting.

No scale-free networks with large values of γ were observed. The reason for this is clear. Indeed, the power-law dependence of the degree distribution can be observed only if there exists for at least 2 or 3 decades of degree. For this, the networks have to be large: their size should be, at least, $t > 10^{2.5(\gamma-1)}$. Then, if γ is large, one practically has no chances to find the scale-free behavior.

In Fig. 5.5, in the log-linear scale, we present the values of the γ exponents of all the networks reported as having power-law degree distributions vs. their sizes (see also Table 5.3). One sees that almost all the plotted points are inside of the region restricted by the lines: $\gamma = 2$, $\log_{10} t \sim 2.5(\gamma - 1)$, and by the logarithm of the size of the largest scale-free network – the World-Wide Web – $\log_{10} t \sim 9$.

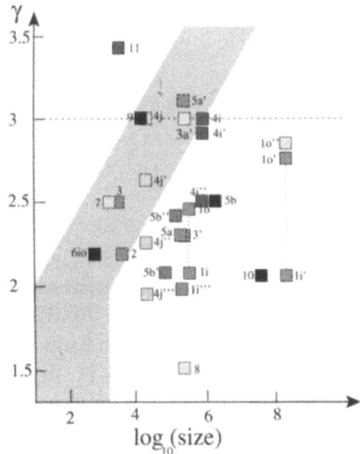

Fig. 5.5. Log-linear plot of the γ exponents of all the networks reported as having power-law (in-, out-) degree distributions (i.e., scale-free networks) vs. their sizes. The line $\gamma \sim 1 + \log_{10} t/2.5$ is the estimate of the finite-size boundary for the observation of the power-law degree distributions for $\gamma > 2$. Here 2.5 is the range of degrees (orders) which we believe is necessary to observe a power law. The dashed line, $\gamma = 3$, is the resilience boundary . This boundary is important for networks which must be stable to random breakdowns. The points are plotted using the data from Table 5.1. The points: $1i$ and $1o$ from complete map of the nd.edu domain of the WWW [52]; $1i'$ and $1o'$ from pages of the WWW scanned by Altavista in October of 1999 [14, 53]; $1o''$ is the γ_o value from another fitting of the same data [32]; $1i'''$ is γ_i for domain level of the WWW in spring 1997 [54]); 2 is γ for the inter-domain level of the Internet in December 1998 [18]; $2'$ is γ for the network of operating AS in one of days in December 1999 [19]; 3 is γ for the router level of the Internet in 1995 [18]; $3'$ is γ for the router level of the Internet in 2000 [15]; $4i$ is γ_i for citations of the ISI database 1981 – June 1997 [23]; $4i'$ is the result of the different fitting of the same data [59]; $4i''$ is another estimate obtained from the same data [6, 11]; $4j$ is γ_i for citations of the Phys. Rev. D **11-50** (1975-1994) [23]; $4j'$ is the different fitting of the same data [59]; $4j''$ is another estimate from the same data [6, 11]; $4j'''$ is γ_i for citations of the Phys. Rev. D (1982-June 1997) [22]; $5a$ is the γ exponent for the collaboration network of movie actors [4]; $5a'$ is the result of another fitting for the same data [8]; $5b$ is γ for the collaboration network of MEDLINE [25]; $5b'$ is γ for the collaboration net collected from mathematical journals [26]; $5b''$ is γ for the collaboration net collected from neuro-science journals [26]; $6io$ is $\gamma_i = \gamma_o$ for networks of metabolic reactions [16]; 7 is γ of the network of protein-protein interactions (yeast proteome) [27, 57]; 8 is γ from word web in the range below the crossover point [49]; 9 is γ of large digital electronic circuits [55]; 10 is γ_i of the telephone call graph [58]; 11 is γ from web of human sexual contacts [56].

Table 5.3. Sizes and values of the γ exponent of the networks or subgraphs reported as having power-law (in-, out-) degree distributions. For each network (or class of networks) data are presented in more or less historical order, so that the recent exciting progress is visible. Errors are not shown. They depend on the size of a network and on the value of γ. [1]The data for the network of operating AS was obtained for one of days in December 1999. [2]The value of the γ exponent was estimated from the degree distribution plot in [15]. [3]The network of protein-protein interaction is treated as undirected. [4]The value of the γ exponent for the word web is given for the range of degrees below the crossover point. [5]The out-degree distribution of the telephone call graph cannot be fitted by a power-law dependence. [6]In fact, the data was collected from a small set of vertices of the web of human sexual contacts. These vertices almost surely have no connections between them. [7]These food webs are truly small

Network or subgraph	# of nodes	# of edges	γ	Ref.
complete map nd.edu domain	$325,729$	$1,469,680$	$\gamma_i = 2.1$ $\gamma_o = 2.45$	[52]
pages of WWW scanned by Altavista in October of 1999	$2.711 \ 10^8$	$2.130 \ 10^9$	$\gamma_i = 2.1$ $\gamma_o = 2.7$	[14, 53]
"———" (another fitting of the same data)			$\gamma_i = 2.10$ $\gamma_o = 2.82$	[32]
domain level of the WWW in spring 1997	$2.60 \ 10^5$	—	$\gamma_i = 1.94$	[54]
inter-domain level of the Internet in Dec. 98	4389	8256	2.2	[18]
net of operating AS in Internet [1]	6374	13641	2.2	[19]
router level of the Internet in 1995	3888	5012	2.5	[18]
router level of the Internet in 2000 [2]	$\sim 150,000$	$\sim 200,000$	~ 2.3	[15]
citations, ISI database 1981 – June 97	$783,339$	$6,716,198$	$\gamma_i = 3.0$	[23]
"———" (another fitting, same data)			$\gamma_i = 2.9$	[59]
"———" (another estimate, same data)			$\gamma_i = 2.5$	[6, 11]
citations, Phys. Rev. D **11-50** (1975-94)	$24,296$	$351,872$	$\gamma_i = 3.0$	[23]
"———" (another fitting, same data)			$\gamma_i = 2.6$	[59]
"———" (another estimate, same data)			$\gamma_i = 2.3$	[6, 11]
citations, Phys. Rev. D (1982-June 1997)	—	—	$\gamma_i = 1.9$	[22]
collab. network of movie actors	$212,250$	$61,085,555$	2.3	[4]
"———" (another fitting of the same data)			3.1	[8]
collab. network of MEDLINE	$1,388,989$	$1.028 \ 10^7$	2.5	[25]
collab. net collected from math. jour.	$70,975$	0.132×10^6	2.1	[26]
collab. net collected from neuro-science jour.	$209,293$	1.214×10^6	2.4	[26]
networks of metabolic reactions	$\sim 500 - 800$	~ 2000	$\gamma_i = 2.2$ $\gamma_o = 2.2$	[16]
net of protein-protein interactions (yeast) [3]	1870	2240	~ 2.5	[27, 57]
word web [4]	$470,000$	$17,000,000$	1.5	[49]
digital electronic circuits	2×10^4	4×10^4	3.0	[55]
telephone call graph [5]	47×10^6	8×10^7	$\gamma_i = 2.1$	[58]
web of human sexual contacts [6]	2810	—	3.4	[56]
food webs [7]	$93 - 154$	$405 - 366$	~ 1	[33]

So, many questions can be addressed: What is the nature of power-laws in networks? One may directly relate them to self-organized criticality. While growing under mechanism of preferential linking, networks self-organize into scale-free structures, that is, are in a critical state. This critical state is realized for a wide range of parameters of preferential linking, namely for any linear preference function (more rigorously, for any preference function which is asymptotically linear at large k [6]). The linear growth of networks may produce scale-free structures. Then, one may ask: What degree distributions does the accelerated growth produce?

5.3 General Relations for the Accelerated Growth

Let us start with general considerations and do not restrict ourselves by some specific model. Let the average degree grows as a power of t, $\overline{k} \propto t^a$, that is, the total number of edges $L(t) \propto t^{a+1}$. Here $a > 0$ is the growth exponent. The consideration is valid not only for degree, but also for in-, and out-degrees, so we use the same notation k for all them. The power-law type of acceleration we have chosen since one may hope that it provide scale-free networks. We suppose from the very beginning that this is the case and then check our assumption.

For the accelerated growth, the degree distribution may be non-stationary. It is natural to choose its power-law part in the form

$$P(k,t) \sim t^z k^{-\gamma}. \tag{5.2}$$

Here we have introduced new exponent $z > 0$ [3, 47, 48] (recall that we consider only $a > 0$). This form is valid only in the range $k_0(t) < k < k_{cut}(t)$. Using the normalization condition $\int_{k_0(t)}^{\infty} dk\, t^z k^{-\gamma} \sim 1$ gives

$$k_0(t) \sim t^{z/(\gamma-1)}. \tag{5.3}$$

This estimate is valid for any $\gamma > 1$.

The cutoff $k_{cut}(t)$ is estimated from the condition $t \int_{k_{cut}(t)}^{\infty} dk\, t^z k^{-\gamma} \sim 1$. Therefore,

$$k_{cut}(t) \sim t^{(z+1)/(\gamma-1)} \tag{5.4}$$

(compare with (5.1) for the linear growth.) Equation (5.4) holds for any $\gamma > 1$.

We will consider two cases (see Fig. 5.6), $1 < \gamma < 2$ and $\gamma > 2$. Recall that we do not account for mortality of vertices.

(i) $1 < \gamma < 2$
Recall that the average degree distribution $\overline{k}(t) \sim t^a$. Then

$$t^a \sim \int^{t^{(z+1)/(\gamma-1)}} dk\, k t^z k^{-\gamma} \sim t^{-1+(z+1)/(\gamma-1)}.$$

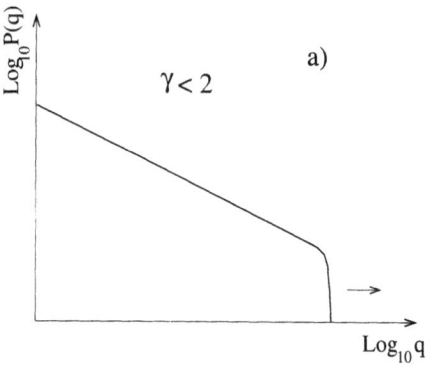

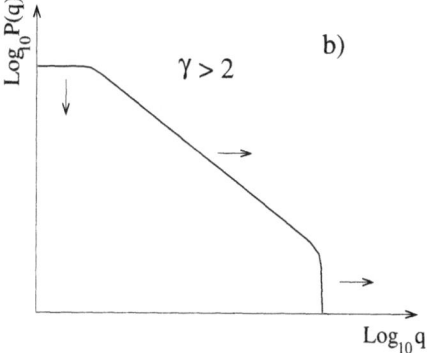

Fig. 5.6. Schematic log-log plots of degree distributions in the two models for accelerating growth of networks. The first model produces the stationary degree distribution with the exponent $\gamma < 2$ **a** at long times. The degree distribution of the second model **b** is non-stationary, $\gamma > 2$. The arrows indicate changes of the distributions as the networks grow

Here the value of the integral is determined by its upper limit. Therefore, $(z + 1)/(\gamma - 1) = a + 1$, and the cutoff is of the order of the total number of edges in the network,

$$k_{cut}(t) \sim t^{a+1} \sim L(t) \,. \tag{5.5}$$

But this is the maximum possible degree in the problem. In this sense, any cutoff of a degree distribution is absent if $\gamma < 2$. From the last relation, we obtain the γ exponent in such a situation,

$$\gamma = 1 + \frac{z+1}{a+1} \,. \tag{5.6}$$

Here, for $\gamma < 2$, one assumes that $z < a$. The lower boundary for γ, namely $\gamma = 1 + 1/(a+1)$ is approached when $z = 0$, that is, when the distribution is stationary.

(ii) $\gamma > 2$

The integral for the average degree is determined by its lower limit

$$t^a \sim \int_{tz/(\gamma-1)} dk \, kt^z k^{-\gamma} \sim t^{z-z(\gamma-2)/(\gamma-1)} \, .$$

Hence

$$\gamma = 1 + \frac{z}{a} \tag{5.7}$$

and $z > a$ to keep $\gamma > 2$. Notice that this relation is not valid for $a = 0$. One sees that, in this case, the degree distribution cannot be stationary: $z > a > 0$.

5.4 Scaling Relations for Accelerated Growth

For simple scale-free networks that grow in a linear mode, simple scaling relations can be written [7, 9]. Let us briefly describe the corresponding scaling relations for the accelerated growth. If vertices in a growing network do not die, one can label them by their "birth date" $0 < s < t$. We denote by $p(k, s, t)$ the probability that the vertex s is of the degree q. The average degree of a vertex s at time t is $\overline{k}(s, t) \equiv \int dk \, kp(k, s, t)$.

For networks that we consider the $\overline{k}(s, t)$ is

$$\overline{k}(s, t) \propto t^{\delta} \left(\frac{s}{t}\right)^{-\beta}, \tag{5.8}$$

where β and γ are scaling exponents. One can show [48] that

$$p(k, s, t) = [1/\overline{k}(s, t)]g[k/\overline{k}(s, t)], \tag{5.9}$$

where $g[\ ]$ is some scaling function, therefore

$$p(k, s, t) = t^{-\delta} \left(\frac{s}{t}\right)^{\beta} g\left[kt^{-\delta} \left(\frac{s}{t}\right)^{\beta}\right]. \tag{5.10}$$

Using the relation $P(k, t) = t^{-1} \int_0^t ds p(k, s, t)$ yields

$$\int_0^{\infty} dx \, t^{-\delta} x^{\beta} g[kt^{-\delta} x^{\beta}] \propto t^{\delta/\beta} k^{-1-1/\beta} \propto t^z k^{-\gamma}, \tag{5.11}$$

whence we obtain relations for the scaling exponents:

$$\gamma = 1 + 1/\beta \tag{5.12}$$

and

$$z = \delta/\beta \, . \tag{5.13}$$

Taking into account these relations gives the scaling form

$$p(k, s, t) = \frac{s^{1/(\gamma-1)}}{t^{(z+1)/(\gamma-1)}} g\left[k\frac{s^{1/(\gamma-1)}}{t^{(z+1)/(\gamma-1)}}\right].$$ (5.14)

Similarly, one can find the scaling form for the degree distribution.

$$P(k, t) = t^z k^{-\gamma} G(kt^{-(1+z)\beta}) = t^z k^{-\gamma} G(kt^{-(1+z)/(\gamma-1)}),$$ (5.15)

where $G(\ )$ is a scaling function. When $z = 0$, (5.14) and (5.15) coincide with the scaling relations [7, 9] for linearly growing networks.

Notice that it is sufficient to know a and only one exponent of $\gamma, \beta, z, \delta,$ or x to find all the others.

5.5 What Are the Degree Distributions Produced by Acceleration?

Let us discuss several illustrative examples. To begin with, we consider a network growing under mechanism of preferential linking, in which the number of new connections increases as a power law in time. At this point we do not discuss the origin of this power-law dependence. Let it be equal to $c_0 t^a$, where c_0 is some positive constant. Here it is convenient to study the in-degree distribution, so that k will be in-degree. In such an event we are interested only in incoming connections, so that the outgoing ends of new edges may be attached to any vertices of the network or even be outside of the net.

Let the probability that a new edge becomes attached to a vertex of in-degree k be proportional to $k + A(t)$, where $A(t)$ is some additional attractiveness of vertices. Two particular cases of this linear preferential linking are considered below in the framework of a simple continuum approach [5, 9, 48].

5.5.1 Model for $\gamma < 2$

If the additional attractiveness is constant, $A = \text{const}$, the continuum equation for the average in-degree $\bar{k}(s, t)$ of individual vertices that born at time s and are observed at time t is of the form

$$\frac{\partial \bar{k}(s, t)}{\partial t} = c_0 t^a \frac{\bar{k}(s, t) + A}{\int_0^t du[\bar{k}(u, t) + A]}$$ (5.16)

with additional starting and boundary conditions $\bar{k}(0, 0) = 0$ and $\bar{k}(t, t) = 0$. Here we supposed that new vertices have no incoming edges. We use this assumption only for brevity. Naturally, the total in-degree of the network is $\int_0^t du\bar{k}(u, t) = c_0 t^{a+1}/(a + 1)$. This also can be seen by integrating both the sides of (5.16) over s. Taking into account the last equality yields the solution of (5.16):

$$\overline{k}(s,t) = A \left(\frac{s}{t}\right)^{-(a+1)}. \tag{5.17}$$

Therefore, β exponent equals $a + 1 > 1$, so that using scaling relation (5.12) gives

$$\gamma = 1 + \frac{1}{a+1} < 2. \tag{5.18}$$

One may also apply the following simple relation of the continuum approach:

$$P(k,t) = \frac{1}{t}\int_0^t ds\, \delta(k - \overline{k}(s,t)) = -\frac{1}{t}\left(\frac{\partial\overline{k}(s,t)}{\partial s}\right)^{-1}\Bigg|_{s=\overline{k}(s,t)}. \tag{5.19}$$

This equality follows from the fact that the solution of the master equation for the probability $p(k,s,t)$ in the continuum approximation is the δ-function. From (5.17) and (5.18) we obtain the in-degree distribution

$$P(k,t) = \frac{A^{1/(a+1)}}{a+1} k^{-[1+1/(a+1)]}, \tag{5.20}$$

which is stationary. We have shown in Sect. 5.3 that when $\gamma = 1 + 1/(a+1)$, the (in-) degree distribution must be stationary, and exponent z is zero. This is the case for the network under consideration.

5.5.2 Model for $\gamma > 2$

Now we choose a different rule of attachment of new edges to vertices. Let the additional attractiveness be time dependent. Furthermore, let it be proportional to the average in-degree of the network, $c_0 t^a/(a+1)$, at the birth of an edge, $A(t) = Bc_0 t^a/(a+1))$. Here $B > 0$ is some constant. Analogously to the above we obtain the non-stationary in-degree distribution

$$P(k,t) \sim t^{a(1+B)/(1-Ba)} k^{-[1+(1+B)/(1-Ba)]} \tag{5.21}$$

for $k \gg t^a$. Hence the γ exponent is

$$\gamma = 1 + \frac{1+B}{1-Ba} > 2. \tag{5.22}$$

The scaling regime is realized when $Ba < 1$.

It is known that the used continuous approach gives exact results for the scaling exponents of the growing networks with a constant density of connections [7]. Nevertheless, it is approximate, so we have checked the obtained above results by simulation.

The results of the simulation of considered models are shown in Figs. 5.8 and 5.9. The size of networks in both studied cases is 10000 sites. The number of the attempts equals 1000. In Fig. 5.8, we present the log-log plots of the average

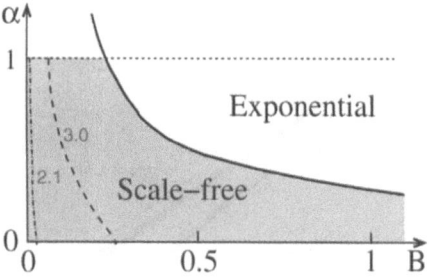

Fig. 5.7. Phase diagram of the networks with the accelerating growth under consideration. The networks are out of the class of scale-free nets ("exponential") above the line $\alpha = 1/B$. The exponent γ equals 3 on the dashed line and 2.1 (value for the World Wide Web) on the dash-dotted one. $\gamma < 2$ on the line $B = 0$

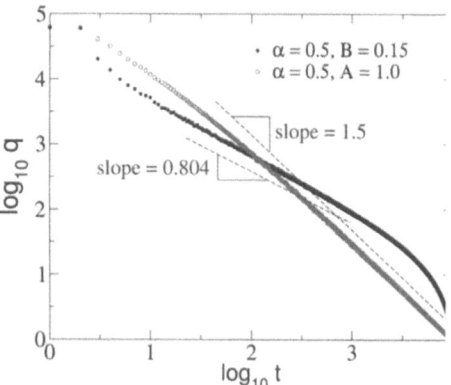

Fig. 5.8. Log-log plot of the average connectivity of a site vs its number (birth time) for the considered models. For the first model, $\alpha = 0.5, n = 1, A = 1.0, c_0 = 1.0$. For the second model, $\alpha = 0.5, n = 1, B = 0.15, c_0 = 1.0$. The *dashed lines* have the slopes equal to the values of the scaling exponent β obtained analytically

connectivity vs number of a site for $\alpha = 0.5, n = 1, A = 1.0, c_0 = 1.0$ (the first model) and for $\alpha = 0.5, n = 1, B = 0.15, c_0 = 1.0$ (the second one). In Fig. 5.9, for these values of parameters of the models, we show the log-log plots of the connectivity distribution.

The obtained values of the scaling exponents are within the error of the simulation from the corresponding ones found analytically. The values $\beta = 1.46$ (1.5) are obtained from the simulation and analytically (in brackets) for the first model with the written out parameters, $\beta = 0.85$ (0.804) are the corresponding values for the second model.

$\gamma = 1.69$ (1.667) and $\gamma = 2.19$ (2.243) are the values of the critical exponent of the connectivity distribution obtained for the first and for the second models, relatively. One may see that the correspondence is really good.

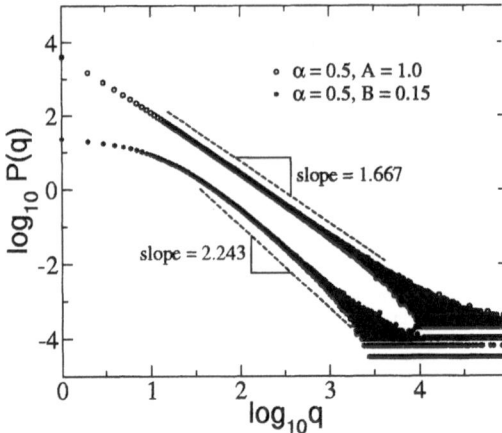

Fig. 5.9. Log-log plot of the distribution of the number of incoming links of sites for the considered models. For the first model, $\alpha = 0.5, n = 1, A = 1.0, c_0 = 1.0$. For the second model, $\alpha = 0.5, n = 1, B = 0.15, c_0 = 1.0$. The dashed lines have the slopes equal to the values of the scaling exponent γ obtained analytically. For better presentation, the dependences are displaced along the vertical axis

Several different values of the scaling exponent of the distribution of incoming links in the World Wide Web were published (as far as we know, any data on the exponent β are absent yet). The available data are $\gamma = 2.1$ [52, 53, 14]. The most huge area was studied in [14], so the value $\gamma = 2.1$ seems to be the best one. As we have noted, one may assume reasonably that α is small in the real networks. We have shown that, in such a situation, the lower boundary for the possible values of γ is slightly below 2. We have demonstrated that, for $\gamma > 2$, the connectivity distribution has to be non-stationary if the growth of the network is accelerating.

There are no data that let us learn whether the connectivity distributions of the World Wide Web and the Internet are stationary or not. Our results make this question intriguing.

The World Wide Web is still in the initial stage of its evolution. Perhaps, the parameters of the accelerating growth will change. In this case, our answers demonstrate the possibility of changing of γ. We have shown that it may become even less than 2 in future.

To demonstrate all the existing possibilities we have considered the models of growing networks with the particular rules of the preferential attachment of new links. These models cover the range of possibilities but provide us only with particular values of the scaling exponents. Of course, there exists a lot of additional factors (aging [9] and mortality [8, 9] of sites, etc.) which may change these particular values.

5.6 One Practical Example: The Word Web

The weak point of network science is the absence of a convincing comparison of numerous schematic models with real networks. Most of models of growing networks only demonstrate intriguing effects but, in fact, are very var from reality. Available empirical data usually can be explained by applying various models with fitting parameters. As a rule, only the exponent of the empirical degree distribution is used for comparison.

Here we consider an exceptional situation, where a reasonable comparison of the model of a growing network with empirical data is possible *without any fitting*. Moreover, it is the idea of the accelerated growth that yields an excellent agreement.

The problem of human language is a matter of immense interest of various sciences. How did language begin? How does language evolve? What is its structure? Quite recently, a novel approach to language was proposed [49]. Human language was considered as a complex network of interacting words. Vertices in this Word Web are distinct words of language, and undirected edges are connections between interacting words.

Words interact when they meet in sentences. Different reasonable definitions yield very similar structures of the Word Web. For example, we can connect the nearest neighbors in sentences. This means that the edge between two words of language exists if these words are the nearest neighbors in at least one sentence in the bank of language. One sees that multiple connections are absent. Of course, this is a rather naive definition, but it is also possible to account for other types of correlations between words in a sentence [49]. The resulting network gives the image of language, which is available for statistical analysis.

The empirical degree distribution [49] of the Word Web is very complex (see Fig. 5.10). Therefore, a perfect description of these data without fitting would be convincing. Indeed, it is hardly possible to describe such a complex form of the distribution completely by coincidence. We show below that a minimal model of the evolving Word Web [50], with only known parameters of this network, provides such a perfect description.

In [49], the Word Web was constructed after processing 3/4 million words of the British National Corpus. The British Corpus is a collection of text samples of both spoken and written modern British English. The resulting network contains about 470 000 vertices. The average degree is $\bar{k} \approx 72$. These are the only parameters of the network we know and can use in the model.

Notice that the quality of the empirical data is [49] is high: the range of degrees is five decades. The empirical degree distribution has two power-law regions with exponents 1.5 and about 3 (the latter value is less precise, since statistics in this region is worse). The crossover point and the cutoff due to finite-size effect can be easily indicated (see Fig. 5.10).

We treat language as a growing network of interacting words. At its birth, a new word already interacts with several old ones. New interactions between old words emerge from time to time, and new edges emerge. All the time a word

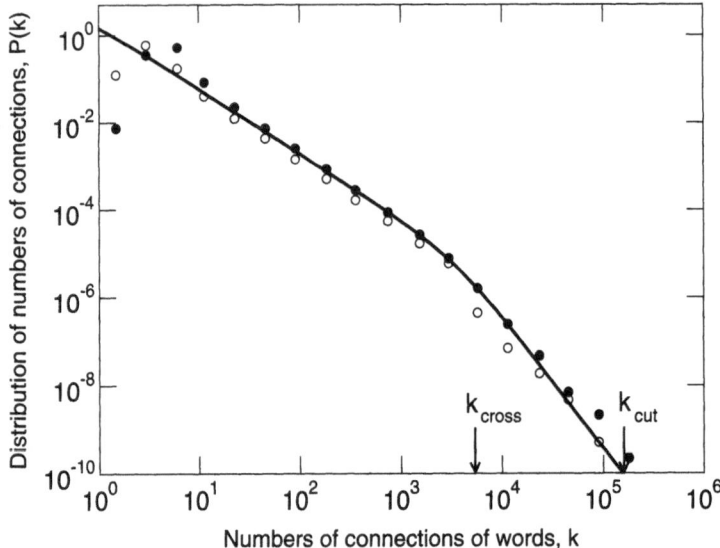

Fig. 5.10. Empirical degree distribution of the Word Web (points) [49]. Empty and filled circles correspond to different definitions of the interactions between words in sentences. The solid line [50] shows the result of our calculations using the known parameters of the Word Web, namely the size $t \approx 470\,000$ and the average number of connections, $\bar{k}(t) \approx 72$. The arrows indicate the theoretically obtained point of crossover, k_{cross}, between the regions with exponents $3/2$ and 3, and the cutoff k_{cut} of the power-law dependence due to finite-size effect

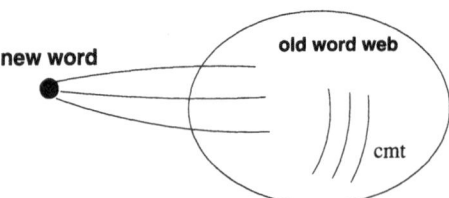

Fig. 5.11. Scheme of the Word Web growth. At each time step a new word emerge, so that t is the total number of words. It connects to $m \sim 1$ preferentially chosen old words. Simultaneously cmt new edges emerge between pairs of preferentially chosen old words. We use the simplest rule of the preferential attachment when a node is chosen with the probability proportional to the number of its connections

lives, it enters in new "collaborations". Therefore the number of connections grows more rapidly than the number of words: the growth of the Word Web is accelerated.

How do words find their collaborators in language? Here we again use the idea of preferential linking [4], again the principle *"popularity is attractive"* works.

We use the following rules of the network growth (see Fig. 5.11) [50].

(1) At each time step, a new vertex (word) is added to the network, and the total number of words is t.

(2) At its birth, a new word connects to several old ons. Let, in average, this number be m, so that this number is not necessary integer. We use the simplest natural version of preferential linking: a new word become connected with some old one i with the probability proportional to its degree k_i, like in the Barabasi-Albert model [4].

(3) In addition, cmt new edges emerge between old words, where c is a constant coefficient that characterizes a particular network. If each vertex makes new connections with a constant rate, this linear dependence on time naturally arises. These new edges emerge between old words i and j with the probability proportional to the product of their degrees $k_i k_j$ [10].

These simple rules define the minimal model that can be solved exactly. Here we discuss only the results of the continuum approach. In this case, the approach gives an excellent description of the degree distribution and the proper values of exponents.

In the model that we discuss, words are actually considered as collaborators in language. In our approach the essence of the evolution of language is the evolution of collaborations between words. Therefore the situation for the Word Web should be rather similar to that for networks of collaborations. The equivalent model was applied to scientific collaboration nets [26], but the more complex nature of these networks makes the comparison impossible.

As above, in the continuum approximation, we can write the equation for the average degree at time t of the word that emerged at time s:

$$\frac{\partial \overline{k}(s,t)}{\partial t} = (m + 2cmt) \frac{\overline{k}(s,t)}{\int_0^t du\, \overline{k}(u,t)} , \qquad (5.23)$$

where the initial condition is $\overline{k}(0,0) = 0$ and the boundary one is $\overline{k}(t,t) = m$.

One can see that the total degree of the network is $\int_0^t du\, \overline{k}(u,t) = 2mt + cmt^2$, so that its average degree at time t is equal to $\overline{k}(t) = 2m + cmt$.

The solution of (5.23) is of a singular form

$$\overline{k}(s,t) = m \left(\frac{cmt}{cms} \right)^{1/2} \left(\frac{2m + cmt}{2m + cms} \right)^{3/2} . \qquad (5.24)$$

The form of this equation indicates the presence of two distinct regimes in this problem. Using (5.19) and (5.24) readily yields the non-stationary degree distribution

$$P(k,t) = \frac{1}{ct} \frac{cs(2 + cs)}{1 + cs} \frac{1}{k} , \qquad (5.25)$$

where $s = s(k,t)$ is the solution of (5.24). Notice that, formally speaking, the number m is absent in (5.25). This is the consequence of our definition of the coefficient cm (see above).

From (5.24) and (5.25), one sees that the non-stationary degree distribution has two regions with different behaviors separated by the crossover point

$$k_{cross} \approx m\sqrt{ct}(2 + ct)^{3/2} \,. \qquad (5.26)$$

The crossover moves in the direction of large degrees as the network grows. Below this point, the degree distribution is stationary,

$$P(k) \cong \frac{\sqrt{m}}{2}k^{-3/2} \,. \qquad (5.27)$$

Above the crossover point, we obtain the behavior

$$P(k,t) \cong \frac{(2m + cmt)^3}{4}k^{-3} \,. \qquad (5.28)$$

so that the degree distribution is non-stationary in this region. Thus, we have obtain two distinct values for the degree distribution exponent, namely, $3/2$ and 3.

The model that we consider has two limiting cases. When $c = 0$, it turns to be the Barabási-Albert model, where $\gamma = 3$. When m is small but cm is large, we come to the network from Sect. 5.5.1 which has $\gamma = 3/2$ and a stationary degree distribution. Thus these two values of γ are not surprising. The important point is that the crossover is observable even though $cmt \gg m$.

The degree distribution has one more important point, the cutoff produced by finite-size effect. We estimate its position from the condition $t \int_{k_{cut}}^{\infty} dk P(k,t) \sim 1$ (see Sects. 5.2 and 5.3). This yields

$$k_{cut} \sim \sqrt{\frac{t}{8}}(2m + cmt)^{3/2} \,. \qquad (5.29)$$

Using (5.26) and (5.28) one can estimate the number of words above the crossover:

$$N_c \approx t \int_{k_{cross}}^{\infty} dk P(k,t) \sim \frac{m}{8c} \,. \qquad (5.30)$$

We know only two parameters of the Word Web that was constructed in [49], namely $t = 0.470 \times 10^6$ and $\overline{k}(t) = 72 = 2m + cmt \approx cmt$. About m we know only that it is of the order of 1. From the above relations, one sees that the dependence on m is actually weak and is not noticeable in log-log-scale plots. In fact, m is inessential parameter of the model. Hence we can set its value to 1.

In Fig. 5.10, we plot the degree distribution of the model (the solid line). To obtain the theoretical curve, we used (5.24) and (5.25) with $m = 1$ and $c \approx \overline{k}(t)/t$. A rather inessential deviations from the continuum approximation are accounted for in the small-degree region ($k \sim 10$). One sees that the agreement with the empirical data [49] is fairly good. Note that we do not used any fitting. However, for a better comparison, in Fig. 5.10, the theoretical curve is displaced upwards. Actually, this is not a fitting, since we have to exclude two empirical points with the smallest degrees. These points are dependent on the method of the construction of the Word Web, on specific grammar, so that any comparison in this region is meaningless in principle.

From (5.26) and (5.29), we find the characteristic values for the crossover and cutoff, $k_{cross} \approx 5.1 \times 10^3$, that is, $\log_{10} k_{cross} \approx 3.7$, and $\log_{10} k_{cut} \approx 5.2$. From Fig. 5.10 we see that these values coincide with the experimental ones. We should emphasize that the extent of agreement is truly surprising. The minimal model does not account for numerous, at first sight, important factors, e.g., the death of words, the variations of words during the evolution of language, etc.

The agreement is convincing since it is approached over the whole range of values of k, that is, over five decades. In fact, the Word Web turns out to be very convenient in this respect since the total number of edges in it is extremely high, about 3.4×10^7 edges, and the value of the cutoff degree is large.

Note that few words are in the region above the crossover point $k_{cross} \approx 5.1 \times 10^3$. These words have a different structure of connections than words from the rest part of language. With the growth of language, k_{cross} increases rapidly but, as it follows from (5.30), the total number N_c of words of degree greater than k_{cross} does not change. It is a constant of the order of $m^2/(8cm) \sim 1/(8c) \approx t/(8\overline{k}) \sim 10^3$, that is of the order of the size of a small set of words forming the kernel lexicon of the British English which was estimated as $5,000$ words [51] and is the most important core part of language. Therefore, our concept suggests that the number of words in this part of language does not depend essentially of the size of language. Formally speaking, the size of this core determined by the value of the average rate c with which words find new partners in language.

If our simple theory of the evolution of language is reasonable then the sizes of the cores of primitive languages are close to those for modern "developed" languages.

5.7 Conclusions

The nonlinear growth of networks is a more general situation than the linear growth. In real evolving networks, the nonlinear, in particular, accelerated growth is widespread and is the rule and not the exception. In many cases, it is impossible to understand the nature of an evolving network without accounting for this acceleration.

The complicating circumstance is that existing empirical data clearly indicate the presence of the acceleration but usually fail to yield its quantitative description. Theoreticians may easily choose any functional form for the nonlinear growth, but do these beautiful dependences have any relation to reality?

We have described the possible degree distributions of such networks and have fixed the lower boundary for the scaling exponent γ. Only the power-law time-dependence of the input flow of new links can keep the network inside of the class of scale-free networks. Nevertheless, we have found the region of parameters in which the scale-free structure is impossible. Our results demonstrate the possibility of quite different scenarios for the network evolution and let us hope to approach satisfactory description of the real networks.

The described theory was then extended and applied to the Word Web (evolving networks of interacting words). The key result is the distribution of numbers of connections of words. We have found that the self-organization produces the most connected small kernel lexicon of language, size of which does not change essentially along the language evolution. The degree distribution of words in this core of language crucially differs from the degree distribution of the rest.

Acknowledgments

The author was partially supported by the project POCTI/99/FIS/33141. The author also thanks to S.N. Dorogovtsev, A.V. Goltsev and A.N. Samukhin for the collaboration on these matters.

Electronic address: jfmendes@fc.up.pt

References

1. S.H. Strogatz, Exploring complex networks, Nature **410**, 268 (2001)
2. R. Albert and A.-L. Barabási, Statistical mechanics of complex networks, Rev. Mod. Phys. 74, 47 (2002)
3. S.N. Dorogovtsev and J.F.F. Mendes, Evolution of networks, Adv. Phys. 51, 1079 (2002)
4. A.-L. Barabási and R. Albert, Emergence of scaling in random networks, Science **286**, 509 (1999)
5. A.-L. Barabási, R. Albert, and H. Jeong, Mean-field theory for scale-free random networks, Physica A **272**, 173 (1999)
6. P.L. Krapivsky, S. Redner, and F. Leyvraz, Connectivity of growing random network, Phys. Rev. Lett. **85**, 4629 (2000)
7. S.N. Dorogovtsev, J.F.F. Mendes, and A.N. Samukhin, Structure of growing networks with preferential linking, Phys. Rev. Lett. **85**, 4633 (2000)
8. R. Albert and A.-L. Barabási, Topology of evolving networks: Local events and universality, Phys. Rev. Lett. **85**, 5234 (2000)
9. S.N. Dorogovtsev and J.F.F. Mendes. Evolution of networks with aging of sites, Phys. Rev. E **62**, 1842 (2000)
10. S.N. Dorogovtsev and J.F.F. Mendes, Scaling behaviour of developing and decaying networks, Europhys. Lett. **52**, 33 (2000)
11. P.L. Krapivsky and S. Redner, Organization of growing random networks, Phys. Rev. E **63**, 066123 (2001)
12. P.L.Krapivsky, G.L. Rodgers, and S. Redner, Degree distributions of growing networks, Phys. Rev. Lett. **86**, 5401 (2001)
13. S.N. Dorogovtsev, J.F.F. Mendes, and A.N. Samukhin. Size-dependent degree distribution of a scale-free growing network, Phys. Rev. E **63**, 062101 (2001)

14. A. Broder, R. Kumar, F. Maghoul, P. Raghavan, S. Rajagopalan, R. Stata, A. Tomkins, and J. Wiener, Graph structure of the web, Proceedings of the 9th WWW Conference, Amsterdam, 15–19 May, 2000, 309
15. R. Govindan, and H. Tangmunarunkit, Heuristics for Internet Map Discovery, Proceedings of the 2000 IEEE INFOCOM Conference, Tel Aviv, Israel, March, 2000, 1371; http://citeseer.nj.nec.com/govindan00heuristics.html
16. Jeong, H., Tombor, B., Albert, R., Oltvai, Z. N. and A.-L. Barabási, 2000, The large-scale organization of metabolic networks, Nature, **407**, 651; cond-mat/0010278
17. S.-H. Yook,H. Jeong, and A.-L. Barabási, Modeling the Internet's large-scale topology, cond-mat/0107417
18. M. Faloutsos, P. Faloutsos, and C. Faloutsos, On power-law relationships of the Internet topology, Comput. Commun. Rev. **29**, 251 (1999)
19. R. Pastor-Satorras, A. Vázquez, and A. Vespignani, Dynamical and correlation properties of the Internet, Phys. Rev. Lett. **87**, 258701 (2001)
20. A. Vázquez, R. Pastor-Satorras, and A. Vespignani, Large-scale topological and dynamical properties of Internet, Phys. Rev. E **65**, 066130 (2002)
21. K.-I. Goh, B. Kahng, and D. Kim, Fluctuation-driven dynamics of the Internet topology, Phys. Rev. Lett. **88**, 108701 (2002)
22. A. Vázquez, Statistics of citation networks, cond-mat/0105031
23. S. Redner, How popular is your paper? An empirical study of citation distribution, Eur. Phys. J. B **4**, 131 (1998)
24. L.A.N. Amaral, A. Scala, M. Barthelemy, and H.E. Stanley, Classes of small-world networks, Proc. Nat. Acad. Sci. USA **97**, 11149 (2000)
25. M.E.J. Newman, The structure of scientific collaboration networks, Proc. Nat. Acad. Sci. U.S.A. **98**, 404 (2001)
26. A.-L. Barabási, H. Jeong, Z. Néda, E. Ravasz, A. Schubert, and T. Vicsek, Evolution of the social network of scientific collaborations, cond-mat/0104162
27. Jeong, H., Mason, S. P., Barabási, A.-L. and Oltvai, Z. N., Lethality and centrality in protein networks, Nature, **411**, 41 (2001)
28. H. Jeong, Z. Néda, and A.-L. Barabási, Measuring preferential attachment for evolving networks, cond-mat/0104131
29. S. Bornholdt and H. Ebel, World Wide Web scaling exponent from Simon's 1955 model, Phys. Rev. E **64**, 035104 (2001)
30. S.N. Dorogovtsev, J.F.F. Mendes, and A.N. Samukhin, WWW and Internet models from 1955 till our days and the "*popularity is attractive*" principle, cond-mat/0009090
31. M. Molloy and B. Reed, A critical point for random graphs with a given degree sequence, Random Structures and Algorithms **6**, 161 (1955)
32. M.E.J. Newman, S.H. Strogatz, and D.J. Watts, Random graphs with arbitrary degree distribution and their applications, Phys. Rev. E **64**, 026118 (2001)
33. J.M Montoya, and R. Solé, Simple rules yield complex webs, Nature **404**, 180 (2000)
34. R. Pastor-Satorras and A. Vespignani, Epidemic spreading in scale-free networks, Phys. Rev. Lett. **86**, 3200 (2001)
35. R. Pastor-Satorras and A. Vespignani, Epidemic dynamics and endemic states in complex networks, Phys. Rev. E **63**, 066117 (2001)
36. B. Bollobás, A probabilistic proof of an asymptotic formula for the number of labelled random graphs, Eur. J. Comb. **1**, 311 (1980)

37. P. Erdös and A. Rényi, On random graphs, Publications Mathematicae **6**, 290 (1959)
38. P. Erdös and A. Rényi, On the evolution of random graphs, Publ. Math. Inst. Hung. Acad. Sci. **5**, 17 (1960)
39. H.A. Simon, On a class of skew distribution functions, Biometrika **42**, 425 (1955)
40. H.A. Simon, *Models of Man* (Wiley, New York, 1957)
41. S.N. Dorogovtsev, J.F.F. Mendes, and A.N. Samukhin, Multifractal properties of growing networks, Europhys. Lett. **57**, 334 (2002)
42. A. Vázquez, A. Flammini, A. Maritan, and A. Vespignani, Modeling of protein interaction networks, cond-mat/0108043
43. A.-L. Barabási, E. Ravasz and T. Vicsek, Deterministic scale-free networks, Physica A **299**, 559 (2001)
44. S.N. Dorogovtsev, A.V. Goltsev, and J.F.F. Mendes, Pseudofractal scale-free web, Phys. REv. E **65**, 066122 (2002)
45. S. Jung, S. Kim, and B. Kahng, A geometric fractal growth model for scale free networks, cond-mat/0112361
46. Z. Burda, J.D. Correia, and A. Krzywicki, Statistical ensemble of scale-free random graphs, Phys. Rev. E **64**, 046118 (2001)
47. S.N. Dorogovtsev and J.F.F. Mendes, Effect of the accelerating growth of communications networks on their structure, Phys. Rev. E 63, 025101 (R) (2001)
48. S.N. Dorogovtsev and J.F.F. Mendes, Scaling properties of scale-free evolving networks: Continuous approach. Phys. Rev. E **63**, 056125 (2001)
49. R. Ferrer and R.V. Solé, The small-world of human language, Proc. Roy. Soc. London B **268**, 2261 (2001)
50. S.N. Dorogovtsev and J.F.F. Mendes, Language as an evolving Word Web, Proc. Royal Soc. B 268, **2603** (2001)
51. R. Ferrer and R. Sole, Two regimes in the frequency of words and the origins of complex lexicons: Zipf's law revised, Journal of Quantitative Linguistics (to appear); Working Papers of Santa Fe Institute, 00-12-068 (2000) http://www.santafe.edu/sfi/publications/Abstracts/00-12-068abs.html
52. Albert, R., Jeong, H. and Barabási, A.-L., 1999, The diameter of the world-wide web, *Nature*, **401**, 130; cond-mat/9907038
53. Kumar, R., Raghavan, P., Rajagopalan, S. and Tomkins, A., 1999, Extracting large-scale knowledge bases from the web, *Proceedings of the 25th VLDB Conference, Edinburgh, Scotland, 7–10 September*, 639–650
54. Adamic, L. A. and Huberman, B. A., 2000, Power-law distribution of the World Wide Web, *comment, Science*, **287**, 2115a; cond-mat/0001459
55. Ferrer, R., Janssen, C. and Solé, R., 2001, The topology of technology graphs: Small world patterns in electronic circuits, *Working Papers of Santa Fe Institute*, 01-05-029, http://www.santafe.edu/sfi/publications/Abstracts/01-05-029abs.html
56. Liljeros, F., Edling, C. R., Amaral, L. A. N., Stanley, H. E. and Åberg, Y., 2001, The web of human sexual contacts, *Nature*, **411**, 907; cond-mat/0106507
57. Wagner, A., 2001, The yeast protein interaction network evolves rapidly and contains few redundant duplicate genes, *Working Papers of Santa Fe Institute*, 01-04-022, http://www.santafe.edu/sfi/publications/Abstracts/01-04-022abs.html
58. Aiello, W., Chung, F. and Lu, L., 2000, A Random Graph Model for Massive Graphs, *Proceedings of the Thirty-Second Annual ACM Symposium on Theory of Computing*, 171–180

59. Tsallis, C. and de Albuquerque, M. P., 2000, Are citations of scientific papers a case of nonextensivity?,
    Eur. Phys. J. B, **13**, 777; cond-mat/9903433
60. S. Solomon and M. Levy, Spontaneous scaling emergence in generic stochastic systems, Int. J. Phys. C **7**, 745 (1996)
61. D. Sornette and R. Cont, Convergent multiplicative processes repelled from zero: power laws and truncated power laws, J. Phys. I (France) **7**, 431 (1997)
62. J.P. Bouchaud and M. Mézard, Wealth condensation in a simple model of economy, Physica A **282**, 536 (2000)

6 Optimization in Complex Networks

Ramon Ferrer i Cancho[1] and Ricard V. Solé[1,2]

[1] ICREA-Complex Systems Lab, Universitat Pompeu Fabra (GRIB), Dr Aiguader
80, 08003 Barcelona, Spain
[2] Santa Fe Institute, 1399 Hyde Park Road, Santa Fe, NM 87501, USA

Abstract. Many complex systems can be described in terms of networks of interacting
units. Recent studies have shown that a wide class of both natural and artificial nets
display a surprisingly widespread feature: the presence of highly heterogeneous distributions of links, providing an extraordinary source of robustness against perturbations.
Although most theories concerning the origin of these topologies use growing graphs,
here we show that a simple optimization process can also account for the observed regularities displayed by most complex nets. Using an evolutionary algorithm involving
minimization of link density and average distance, four major types of networks are
encountered: (a) sparse exponential-like networks, (b) sparse scale-free networks, (c)
star networks and (d) highly dense networks, apparently defining three major phases.
These constraints provide a new explanation for scaling of exponent about −3. The
evolutionary consequences of these results are outlined.

6.1 Introduction

Many essential features displayed by complex systems, such as memory, stability and homeostasis emerge from their underlying network structure [26, 14].
Different networks exhibit different features at different levels but most complex
networks are extremely sparse and exhibit the so-called small-world phenomenon
[28]. An inverse measure of sparseness, the so-called network density, is defined
as

$$\rho = \frac{\langle k \rangle}{n-1} \tag{6.1}$$

where n is the number of vertices of the network and $\langle k \rangle$ is its average degree.
For real networks we have $\rho \in [10^{-5}, 10^{-1}]^3$.

It has been shown that a wide range of real networks can be described by
a degree distribution $P(k) \sim k^{-\gamma}\phi(k/\xi)$ where $\phi(k/\xi)$ introduces a cut-off at
some characteristic scale ξ. Three main classes can be defined [2]. (a) When ξ
is very small, $P(k) \sim \phi(k/\xi)$ and thus the link distribution is single-scaled. Typically, this would correspond to exponential or Gaussian distributions; (b) as
ξ grows, a power law with a sharp cut-off is obtained; (c) for large ξ, scale-free
nets are observed. The last two cases have been shown to be widespread and

---

[3] Statistics performed on Table I in [5]

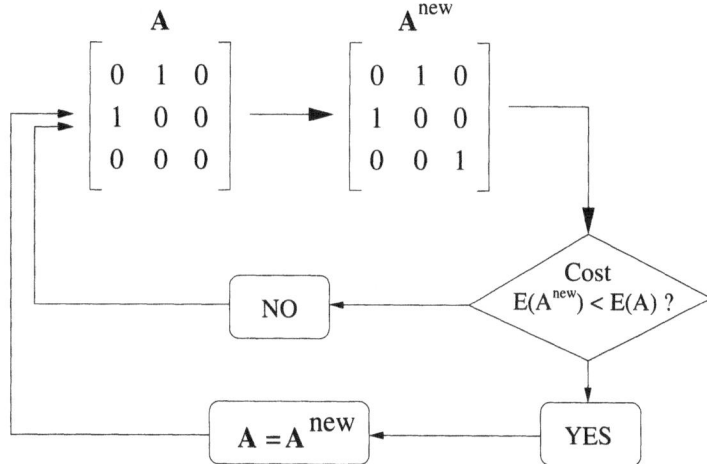

Fig. 6.1. Basic scheme of the minimization algorithm. Starting from a given adjacency matrix **A** the algorithm performs a change in a small number of bits (specifically, with probability ν, each a_{ij} can flip). The energy function e is then evaluated and the new matrix is accepted provided that a lower cost is achieved. Otherwise, we start again with the original matrix. At the beginning, **A** is set up with a fixed density $\rho(0)$ of ones

their topological properties have immediate consequences for network robustness and fragility [5]. The three previous scenarios are observed in: (a) power grid systems and neural networks [2], (b) protein interaction maps [12], metabolic pathways [13] and electronic circuits [16] and (c) Internet topology [13, 8], scientific collaborations [20] and [17] lexical networks.

6.2 Network Optimization

Scale-free nets are particularly relevant due to their extremely high homeostasis against random perturbations and fragility against removal of highly connected nodes[1]. These observations have important consequences, from evolution to therapy [12]. One possible explanation for the origin of the observed distributions would be the presence of some (decentralized) optimization process.

Network optimization is actually known to play a leading role in explaining allometric scaling in biology [29, 7, 3] and has been shown to be a driving force in shaping neural wiring at different scales [9, 18] (see also [6]). In a related context, local and/or global optimization has been also shown to provide remarkable results within the context of channel networks [22]. By using optimality criteria linking energy dissipation and runoff production, the fractal properties in the model channel nets were essentially indistinguishable from those observed in nature. Figure 6.2 displays different optimal transportation networks.

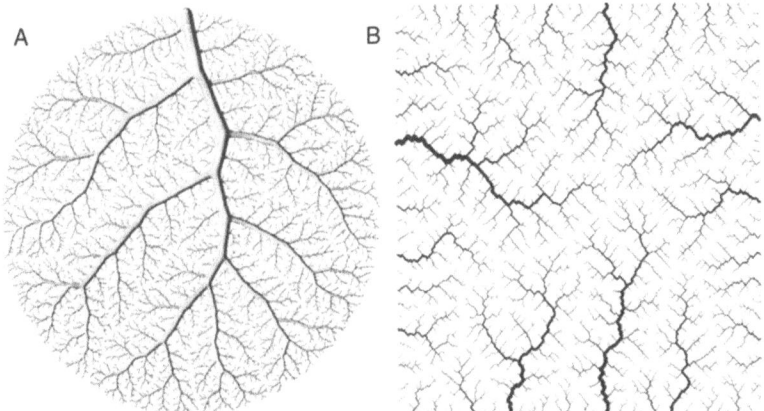

Fig. 6.2. Optimal transport networks in biology (**A**) and geomorphology (**B**). **A.** An optimal tree structure that has been obtained for a vascular system on a two dimensional perfusion area [7]. **B.** An optimal river basin network (also displaying tree structure) that has been generated by minimizing energy expenditure [22]

Several mechanisms of network evolution lead to scale-free structures within the context of complex networks in which the only relevant elements are vertices and connections [4]. Optimization has not been found to be one of them [5]. In this context, it was shown that (Metropolis-based) minimization of both vertex-vertex distance and link length (*i.e.* Euclidean distance between vertices)[15] can lead to the small-world phenomenon and hub formation. This view takes into account Euclidean distance between vertices. Here we show how minimizing both vertex-vertex distance and the number of links leads (under certain conditions) to the different types of network topologies depending on the weight given to each constraint. These two constraints include two relevant aspects of network performance: the cost of physical links between units and communication speed among them.

6.3 The Optimization Algorithm

For the sake of simplicity, we take an undirected graph having a fixed number of nodes n and links defined by a binary adjacency matrix $A = \{a_{ij}\}$, $1 \leq i, j \leq n$. Given a pair of vertices i and j, $a_{ij} = 1$ if they are linked ($a_{ij} = 0$ otherwise) and D_{ij} is the minimum distance between them. At time $t = 0$, we have a randomly wired graph (i.e. a Poisson degree distribution) in which two given nodes are connected with some probability p. The energy function of our optimization algorithm is defined as

$$E(\lambda) = \lambda d + (1 - \lambda)\rho$$

where $0 \leq \lambda, d, \rho \leq 1$. λ is a parameter controlling the linear combination of d and ρ. The normalized number of links, ρ is defined in terms of a_{ij} as

$$\rho = \frac{1}{\binom{n}{2}} \sum_{i<j} a_{ij}$$

and it is equivalent to (6.1). The normalized vertex-vertex distance, d, is defined as $d = D/D^{linear}$ being

$$D = \frac{1}{\binom{n}{2}} \sum_{i<j} D_{ij}$$

the average minimum vertex-vertex distance and $D^{linear} = (n+1)/3$ the maximum value of D that can be achieved by a connected network, that is, that of a linear graph (see Appendix). We define a linear graph as a graph having 2 vertices with degree 1 and $n-2$ vertices with degree 2^4. A graph whose adjacency matrix satisfies

$$a_{ij} = \begin{cases} 1 \text{ if } |i-j| = 1 \\ 0 \text{ otherwise} \end{cases} \tag{6.2}$$

is a linear graph. Such a graph has the maximum average vertex-vertex distance that can be achieved by a connected graph of order n (see Appendix).

The minimization of $E(\lambda)$ involves the simultaneous minimization of distance and number of links (which is associated to cost). Notice that minimizing $E(\lambda)$ implies connectedness (*i.e.* finite vertex-vertex distance) except for $\lambda = 0$, where it will be explicitly enforced.

The minimization algorithm proceeds as follows. At time $t = 0$, the network is set up with a density $\rho(0)$ following a Poissonian distribution of degrees (connectedness is enforced). At time $t > 0$, the graph is modified by randomly changing the state of some pairs of vertices. Specifically, with probability ν, each a_{ij} can switch from 0 to 1 or from 1 to 0. The new adjacency matrix is accepted if $E(\lambda, t+1) < E(\lambda, t)$. Otherwise, we try again with a different set of changes. The algorithm stops when the modifications on $A(t)$ are not accepted T times in a row. The minimization algorithm is a simulated annealing at zero temperature. Figure 6.1 describes the minimization algorithm. Hereafter, $n = 100^5$, $T = \binom{n}{2}^6$, $2/\binom{n}{2}^7$ and $\rho(0) = 0.2$.

---

[4] It can be easily shown through induction on n that such a graph is connected and has no cycles.

[5] Higher values of n were very time consuming. The critical part of the algorithm is the calculation of d which has cost $\Theta(n\rho\binom{n}{2})$, that is, $\Omega(n^2)$ and $O(n^3)$. Faster calculation implies performing an estimation of d on a random subset of vertices or 1st and 2nd neighbors [21] that happened to be misleading.

[6] Intended for expecting that every pair of vertices has been allowed to change its state at least once.

[7] We define the number of changes in the adjacency matrix between generations as $c = |\{a_{ij}(t+1)|i < j \text{ and } a_{ij}(t+1) \neq a_{ij}(t)\}|$. Let $d(t)$ and $\rho(t)$ be respectively the distance and the density at time t. If $c = 1$ then $d(t+1) < d(t)$ and $\rho(t+1) = \rho(t)$ is impossible. If $c > 1$ then $d(t+1) < d(t)$ with $\rho(t+1) = \rho(t)$ is allowed. Thereafter ν is set to enforce $E[c] = 2 > 1$.

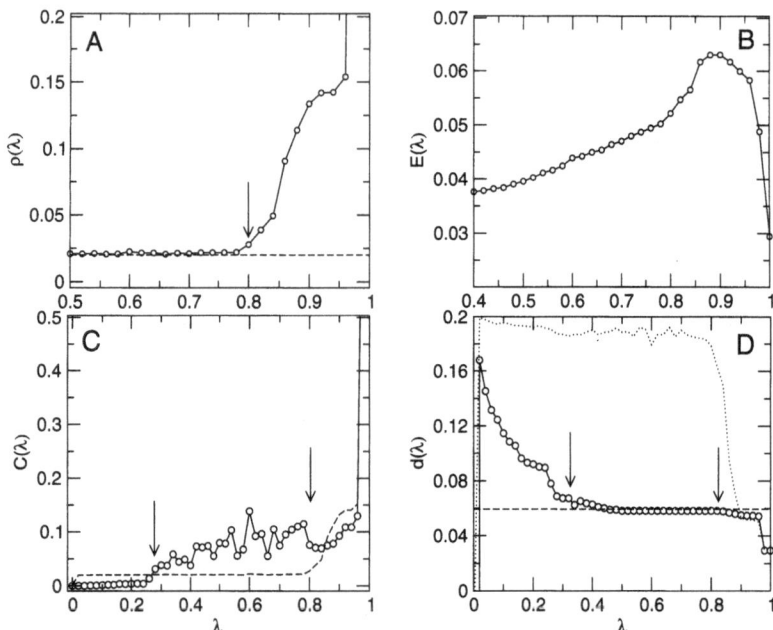

Fig. 6.3. Density (**A**), energy (**B**), clustering coefficient (**C**) and distance (**D**) as a function of λ. Averages over 50 optimized networks with $n = 100$, $T = \binom{n}{2}$, $\nu = 2/\binom{n}{2}$ and $\rho(0) = 0.2$ are shown. **A**: the optimal network becomes a complete graph for λ close to 1. The density of a star network, $\rho_{star} = 2/n = 0.02$ is shown as reference (*dashed line*). The clustering coefficient of a Poissonian network $C_{random} = \langle k \rangle/(n-1)$ is shown as reference in **C**. Notice that $C_{random} \approx \rho$. The normalized distance of a star network is (see Appendix), $d_{star} = 6(n-1)/(n(n+1)) = 0.058$ (*dashed line*) and that of a Poissonian network, $d_{random} = \log n/\log \langle k \rangle$ (*dotted line*) are shown for reference in **D**

We define the degree entropy on a certain value of λ as

$$H(\{p_k\}) = -\sum_{k=1}^{n-1} p_k \log p_k$$

where p_k is the frequency of vertices having degree k and $\sum_{k=1}^{n-1} p_k = 1$. This type of informational entropy will be used in our characterization of the different phases[8].

Some of the basic average properties displayed by the optimized nets are shown against λ in Fig. 6.3. These plots, together with the degree entropy in Fig. 6.4 suggest that four phases are present, separated by three sharp transitions at $\lambda_1^* \approx 0.25$, $\lambda_2^* \approx 0.80$ and $\lambda_3^* \approx 0.95$ (see arrows in Fig. 6.3). The second one

---

[8] Entropy measures of this type have been used in characterizing optimal channel networks and other models of complex systems (see [24]) although they are typically averaged over time.

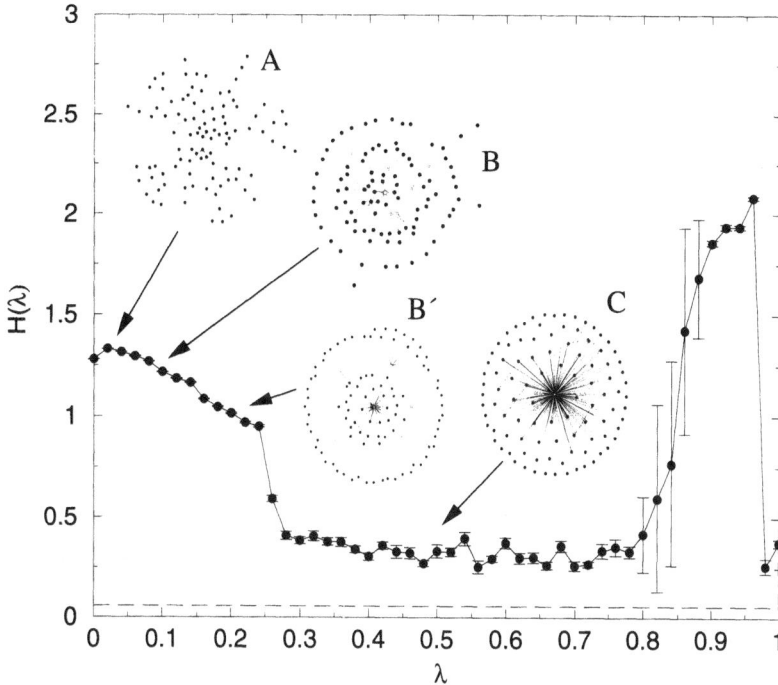

Fig. 6.4. Average (over 50 replicas) degree entropy as a function of λ with $n = 100$, $T = \binom{n}{2}$, $\nu = 2/\binom{n}{2}$ and $\rho(0) = 0.2$. Optimal networks for selected values of λ are plotted. The entropy of a star network, $H_{star} = \log n - [(n-1)/n]\log(n-1) = 0.056$ is provided as reference (dashed line). **A:** an exponential-like network with $\lambda = 0.01$. **B:** A scale-free network with $\lambda = 0.08$. Hubs involving multiple connections and a dominance of nodes with one connection can be seen. **C:** a star network with $\lambda = 0.5$. **B':** a intermediate graph between **B** and **C** in which many hubs can be identified

separates sparse nets from dense nets and fluctuations in $H(\lambda_3^*)$ are specially high. $\rho(\lambda), C(\lambda) \approx 1$ for $\lambda > \lambda_3^* \approx 0.95$. For $\lambda = 0$ and $\lambda = 1$ a Poissonian and a complete ($\rho(\lambda) = 1$) network are predicted, respectively.

6.4 Optimal Degree Distributions

When taking a more careful look at the sparse domain $(0, \lambda_2^*)$, three non-trivial types of networks are obtained as λ grows:

a. Exponential networks, i. e. $P_k \sim e^{-k/\xi}$.
b. Truncated scale-free networks, i. e. $P_k \sim k^{-\gamma}e^{-k/\xi}$ with $\gamma = 3.0$ and $\xi \approx 20$ (for $n = 100$).
c. Star network phase ($\lambda_1^* < \lambda < \lambda_2^*$) *i.e.* a central vertex to which the rest of the vertices are connected to (no other connections are possible). Here,

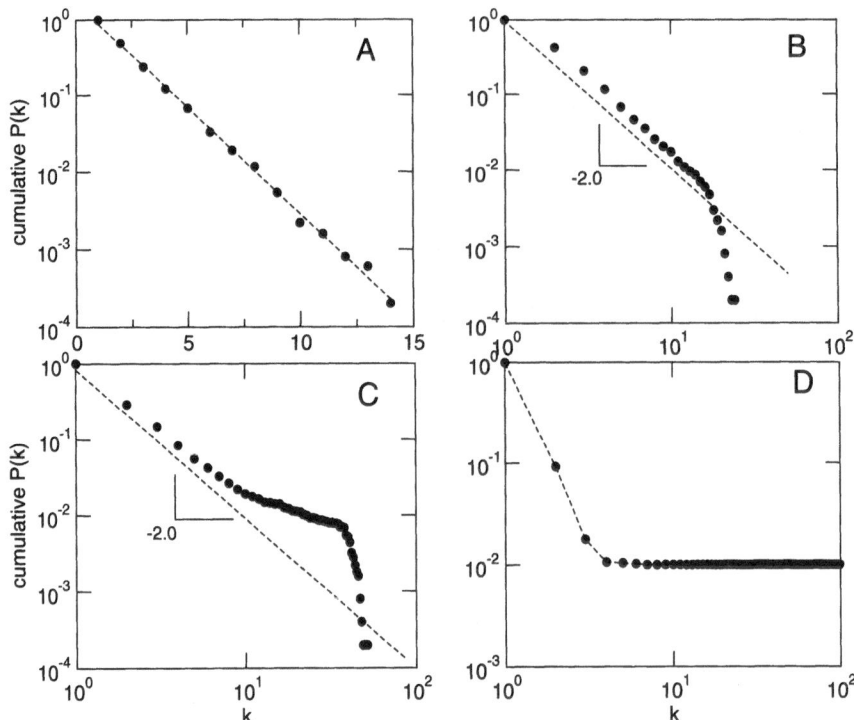

Fig. 6.5. Selected cumulative degree distributions of networks obtained minimizing $E(\lambda)$. Every distribution is an average over 50 optimized networks with $n = 100$, $T = \binom{n}{2}$, $\nu = 2/\binom{n}{2}$ and $\rho(0) = 0.2$. **A:** an exponential-like distribution for $\lambda = 0.01$. **B:** a power distribution with exponent $\gamma = 2.0$ for $\lambda = 0.08$ (with a sharp cutoff at $\xi \approx 20$). **C:** $\lambda = 0.20$. **D:** $\lambda = 0.50$ (almost an star graph)

$$p_k = \frac{n-1}{n}\delta_{k,1} + \frac{1}{n}\delta_{k,n-1} \qquad (6.3)$$

A star graph has the shortest vertex-vertex distance between vertices among all the graphs having a minimal amount of links (see Appendix). Non-minimal densities can be compensated with a decrease in distance, so pure star networks are not generally obtained.

The distributions of (a-c) types and that of a dense network are shown in Fig. 6.5. A detailed examination of the transition between degree distributions reveals that hub formation explains the emergence of (b) from (a), hub competition (b') precedes the emergence of a central vertex in (c). The emergence of dense graphs from (c) consists of a progressive increase in the average degree of non-central vertices and a sudden loss of the central vertex. The transition to the star net phase is sharp. Figure 6.4 shows $\langle H(\lambda)\rangle$ along with plots of the major types of networks. It can be seen that scale-free networks (b) are found close to λ_1^*. The cumulative exponent of such scale-free networks is two and thus

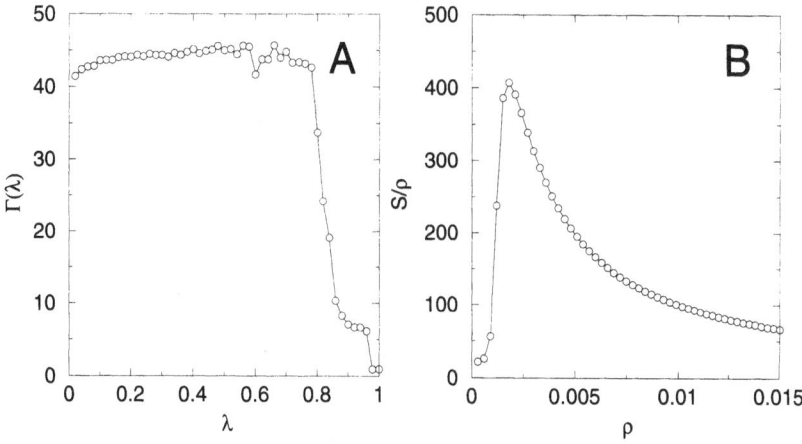

Fig. 6.6. A. The function $\Gamma(\lambda) = (1 - d(\lambda))/\rho(\lambda)$ for the minimum energy configurations. **B.** The cost function S/ρ versus ρ for the Poissonian model

$\gamma = 3.0$, the same that it would be expected for a random network generated with the Barabási-Albert model [4].

Our scenario suggests that preferential attachment networks might emerge at the boundary between random attachment networks (a) and forced attachment (*i.e.* every vertex connected to a central vertex) networks (c) and points that optimization can explain the selection of preferential attachment strategies in real complex networks. In our study, exponential-like distributions appear when distance is minimized under high density pressure, in agreement with the study by Amaral and co-workers on classes of small-world networks [2]. This might be the case of the power grid and of neural networks [2]. If linking cost decreases sufficiently, cliquishness becomes an affordable strategy for reducing vertex-vertex distance. Consistently, graphs tend to a complete graph for high values of λ. The Watts model [28] is a non-trivial example of what cliquishness (i.e. high clustering) can do for smallwordness. High clustering favours small-worldness (as seen for $\lambda \geq \lambda_2^*$) but it is not the only mechanism [10].

We have seen the different optimal topologies depending on the value of λ. We are aimed at defining an absolute measure of optimality depending on λ we can use for ranking the different topologies. We define

$$\Gamma(\lambda) = \frac{1 - d(\lambda)}{\rho(\lambda)} \tag{6.4}$$

as such measure (Fig. 6.6A). A sharp transition from sparse to dense networks is clearly observed for $\lambda \approx 0.8$. According to Fig. 6.6A, the topology ranking becomes,

1. Star networks.
2. Scale-free networks.

3. Exponential networks.
4. Dense networks.

See the Appendix section for a summary of the basic features of the trivial topologies appearing in our study.

A simpler version of the previous scenario appears in the context of Poissonian graphs, where we define the optimality measure as S/ρ, where S is the number of vertices of the largest connected component and ρ is both the expected network density and the probability that a random pair of vertices are linked. Again, the maximum divides networks into disconnected networks and connected networks at high link expense (Fig. 6.6B). $\rho \approx 0.8$ divides low cost strategies from high cost strategies as $\lambda = 0.8$ does in Fig. 6.6A. Notice that the transition is smooth for the former and sharp for the latter. The Poissonian scenario shows the optimization principles that may guide networks in early stages to remain close to the connectedness transition. Once enough connectedness is achieved, networks may be guided by (6.4) or particular values of λ depending on the system.

6.5 Discussion

The network previous results and our conjecture concerning optimization in complex nets requires explaining why star graphs are not found in nature. Different constraints can be restricting the access of star graphs to real systems. Let us list some of them:

- Randomness. The evolution of the topology as λ grows suggests a transition from disorder (exponential degree distribution) to order (star degree distribution).
- Diversity. The number of different star graphs that can be formed with n vertices is n whereas it explodes for exponential and power distributions.
- Robustness. Removing the central hub leaves $n - 1$ connected components, which is the worst case situation.

Whether or not optimization plays a key role in shaping the evolution of complex networks, both natural and artificial, is an important question. Different mechanisms have been suggested to explain the emergence of the striking features displayed by complex networks. Most mechanisms rely on preferential attachment-related rules, but other scenarios have also been suggested [25, 27] in which external parameters have to be tuned. When dealing with biological networks, the interplay between emergent properties derived from network growth and selection pressures has to be taken into account. As an example, metabolic networks seem to result from an evolutionary history in which both preferential attachment and optimization are present. The topology displayed by metabolic networks is scale-free, and the underlying evolutionary history of these nets suggests that preferential attachment might have been involved [11]. Early in the evolution of life, metabolic nets grew by adding new metabolites, and the most connected are actually known to be the oldest ones. On the other hand, several studies have revealed that metabolic pathways have been optimized through evolution in a number of ways. This suggests that the resulting networks are the outcome of both contributions, plus some additional constraints imposed by the available components to the evolving network [19, 23]. In this sense, selective pressures might work by tuning underlying rules of net construction. This view corresponds to Kauffman's suggestion that evolution would operate by taking advantage of some robust, generic mechanisms of structure formation [14].

Appendix

Throughout this paper, different trivial topologies have appeared. Table 6.1 summarizes their features indicating the value of λ at which they appear. Although this paper is concerned with what happens for $\lambda \geq 0$, notice that the linear graph is the expected outcome for $\lambda < 0$, since it implies distance maximization and density minimization. The remaining of this section is devoted to proof that a linear graph and a star graph have the maximum finite distance and the minimum distance (with the constraints of connectedness and having the smallest amount of edges).

A linear graph is a graph having the maximum finite distance or in other words, it is the connected graph having the maximum distance. We will proof it through induction on n. For $n = 2$, there is only one possible connected graph, which trivially has the maximum distance. All linear graphs having the same amount of vertices have the same average vertex-vertex distance. If the graph in (6.2) has the maximum distance for n vertices, will it still be the longest for $n + 1$ vertices? Assuming that the graph in (6.2) is the longest for n vertices, the longest graph of $n + 1$ vertices has to be formed by the longest graph of order n and a new a vertex linked to one of the n existing vertices. Here we define the total vertex-vertex distance as

$$D_n = \sum_{i<j} D_n(i,j) \tag{6.5}$$

Table 6.1. Different trivial topologies with density (i.e. normalized amount of links) ρ, average vertex-vertex distance D, clustering coefficient C, degree distribution entropy H and the values of λ where they are optimal. $-$ indicates absence of known analytical result

Topology	ρ	D	C	H	λ
Poisson	ρ	$\approx \frac{\log n}{\log(\rho(n-1))}$	ρ	$-$	0
Star	$2/n$	$\frac{2(n-1)}{n}$	0	$\log n - \frac{(n-1)}{n}\log(n-1)$	$-$
Complete	1	1	1	0	1
Linear	$2/n$	$\frac{n+1}{3}$	0	$\frac{1}{n}((n-2)\log(n-2) + 2\log2)$ $-\log n$	$\lambda < 0$

where $D_n(i, j)$ is the minimum distance from the i-th vertex to the j-th vertex. We define the average vertex vertex distance as

$$< D_n >= D_n / \binom{n}{2}$$

If D_{n+1}^k is the contribution to D_{n+1} when the new vertex is linked to the k-th existing vertex, $1 \le k \le n$, such an $n + 1$-vertex graph obeys

$$D_{n+1} = D_n + D_{n+1}^k \tag{6.6}$$

where

$$D_{n+1}^k = \sum_{i=1}^{k} i + \sum_{i=2}^{n-k+1} i$$

Previous equation leads to

$$D_{n+1}^k = \binom{n+1}{2} \tag{6.7}$$

for $k = 1$ and $k = n$. In general,

$$D_{n+1}^k = k^2 - (n+1)k + \frac{n^2 + 3n}{2}$$

D_{n+1}^k has one single non-assymptotical minimum (at $k^* = (n-1)/2$) and no non-assymptotical maximum so D_n^k is maximal for $k = 1$ and $k = n$ and $1 \le k \le n$. $k = 1$ or $k = n$ correspond to a graph order $n + 1$ satisfying (6.2), as we wanted to proof.

Substituting (6.7) into (6.6), we get the longest graph of order n satisfys

$$D_n = D_{n-1} + \binom{n}{2}$$

Expanding the previous recursion we get

$$D_n = \sum_{i=2}^{n} \binom{n}{2} = \frac{1}{2} \left(\sum_{i=1}^{n} i^2 - \sum_{i=1}^{n} i \right)$$

After some algebra we have $D_n = n(n^2 - 1)/6$ and thus $< D_n >= (n+1)/3$

It can also be shown through induction on n that a star graph with a degree distribution

$$p_k = \frac{n-1}{n} \delta_{k,1} + \frac{1}{n} \delta_{k,n-1} \tag{6.8}$$

has the minimum distance possible among all possible graphs having $n - 1$ links. For $n = 2$, the only connected graph (and thus the only with finite distance) trivially is the best one having $n-1$ links. If we assume that the graph described in (6.8) is the optimal for n vertices, the optimal graph of $n+1$ vertices has $d_{n+1} = d_n + \Delta^k_{n+1}$ where Δ^k_{n+1} is the contribution to D_{n+1} of the new vertex when linked to the k-th existing vertex. Thereafter, $\Delta^1_{n+1} = 2n - 1$ and $\Delta^k_{n+1} = 3(n-1)$ for $1 < k \leq n$. $\Delta^1_{n+1} < \Delta^{k>1}_{n+1}$ holds for $n > 2$, so the graph of order $n+1$ obeying (6.8) is also the best one with $n-1$ links.

Acknowledgments

We acknowledge R. Pastor-Satorras for helpful discussions and the technical assistance of F. Busquets. We thank M. Magnasco for helpful comments. This work was supported by (and started at) the Santa Fe Institute (RFC and RVS) and grants of the Generalitat de Catalunya (FI/2000-00393, RFC) and the CICYT (PB97-0693, RVS).

References

1. R. Albert, H. Jeong, and A.-L. Barabási: Error and attack tolerance of complex networks. *Nature* 406, 378–381 (2000)
2. L.A. Nunes Amaral, A. Scala, M. Barthélémy, and H. Eugene Stanley: Classes of behaviour of small-world networks. *Proc. Natl. Acad. Sci.* 97, 11149–11152 (2000)
3. J.R. Banavar, A.Maritan and A. Rinaldo: Size and form in efficient transportation networks. *Nature* 399,130–132 (1999)
4. A.-L. Barabási and R. Albert: Emergence of scaling in random networks. *Science* 286, 509–511 (1999)
5. A.-L. Barabási and R. Albert: Statistical mechanics of complex networks. *Reviews of modern physics* 74, 47–97 (2002)
6. S. Bornholdt and K. Sneppen: Robustness as an evolutionary principle. *Proc. R. Soc. Lond. B* 267, 2281–2286 (2000)
7. J. H. Brown and G.B. West, eds: *Scaling in Biology.* (Oxford U. Press, New York, 2000)
8. G. Caldarelli, R. Marchetti, and L. Pietronero: The fractal properties of Internet. *Europhys. Lett.* 52, 386–391 (2000)
9. C. Cherniak: Neural component placement. *Trends Neurosci.* 18, 522–527 (1995)
10. S.N. Dorogovtsev and J.F.F. Mendes: Evolution of random networks. *Adv. Phys.* 51, 1079–1187 (2002)
11. D. Fell and A. Wagner: The small-world of metabolism. *Nature Biotech.* 18, 1121–1122 (2000)
12. H. Jeong, S.P. Mason, A.-L. Barabási, and Z.N. Oltvai: Lethality and centrality in protein networks. *Nature* 411, 41–42 (2001)
13. H. Jeong, B. Tombor, R. Albert, Z.N. Oltvai, and A.-L. Barabási: The large-scale organization of metabolic networks. *Nature* 407, 651–654 (2000)
14. S.A. Kauffman: *The Origins of Order: Self-Organization.* (Oxford University Press, New York, 1993)
15. N. Mathias and V. Gopal: Small worlds: How and why. *Phys. Rev. E,* 63, 021117-021128 (2001)

16. R Ferrer i Cancho, C. Janssen, and R.V. Solé: Topology of technology graphs: small world patterns in electronic circuits. *Phys. Rev. E* 64, 046119 (2001)
17. Ramon Ferrer i Cancho and Ricard V. Solé: The small-world of human language. *Proc. R. Soc. Lond. B* 268, 2261–2266 (2001)
18. G. Mitchinson: Neural branching patterns and the economy of cortical wiring. *Proc. R. Soc. London B* 245, 151–158 (1991)
19. H.J. Morowitz, J.D. Kostelnik, J. Yang, and G.D. Cody: The origin of intermediary metabolism. *Proc. Natl. Acad. Sci. Sci. USA* 97, 7704–7708 (2000)
20. M.E.J. Newman: The structure of scientific collaboration networks. *Proc. Natl. Acad. Sci.* 98, 404–409 (2001)
21. M.E.J. Newman, S.H. Strogatz, and D.J. Watts: Random graphs with arbitrary degree distribution and their applications. *Phys. Rev. E* 64, 026118 (2001)
22. I. Rodriguez-Iturbe and A. Rinaldo: *Fractal River Basins.* (Cambridge U. Press, Cambridge, 1997)
23. P. Schuster: Taming combinatorial explosion. *Proc. Natl. Acad. Sci. Sci. USA* 97, 7678–7680 (2001)
24. R.V. Solé and O. Miramontes: Information at the edge of chaos in fluid neural networks. *Physica D* 80, 171–180 (1995)
25. R.V. Solé, R. Pastor-Satorras, E. Smith, and T. Kepler: A model of large-scale proteome evolution. *Adv. Complex Syst.* 5, 43–54 (2002)
26. S.H. Strogatz: Exploring complex networks. *Nature* 410, 268–276 (2001)
27. A. Vázquez, A. Flammini, A. Maritan, and A. Vespignani: Modeling of protein interaction networks. *Complexus* 1,38–44 (2003)
28. D.J. Watts and S.H. Strogatz: Collective dynamics of 'small-world' networks. *Nature* 393,440–442 (1998)
29. G.B. West, J.H. Brown, and B.J. Enquist: A general model for the origin of allometric scaling laws in biology. *Science* 276, 122–126 (1997)

7 Epidemic Spreading in Complex Networks with Degree Correlations

Marián Boguñá[1], Romualdo Pastor-Satorras[2], and Alessandro Vespignani[3]

[1] Department de Física Fonamental, Universitat de Barcelona, Av. Diagonal 647, 08028 Barcelona, Spain
[2] Department de Física i Enginyeria Nuclear, Universitat Politècnica de Catalunya, Campus Nord, 08034 Barcelona, Spain
[3] Laboratoire de Physique Théorique (UMR du CNRS 8627), Bâtiment 210 Université de Paris-Sud 91405 ORSAY Cedex, France

Abstract. We review the behavior of epidemic spreading on complex networks in which there are explicit correlations among the degrees of connected vertices.

7.1 Introduction

Complex networks arising in the modeling of many social, natural, and technological systems are often growing and self-organizing objects characterized by peculiar topological properties [1, 2]. Many empirical evidences have prompted that most of the times the resulting network's topology exhibits emergent phenomena which cannot be explained by merely extrapolating the local properties of their constituents. Among these phenomena, two of them appear ubiquitous in growing networks. The first one concerns the *small-world* property [3], that is defined by an average path length—average distance between any pair of vertices—increasing very slowly (usually logarithmically) with the network size, N. The second one finds its manifestation in the *scale-free* (SF) degree distribution [1]. This implies that the probability $P(k)$ that a vertex has degree k—it is connected to k other vertices—is characterized by a power-law behavior $P(k) \sim k^{-\gamma}$, where $2 < \gamma \leq 3$ is a characteristic exponent.

The statistical physics approach [1, 2] has been proved a very valuable tool for the understanding and modeling of these emergent phenomena in growing networks and has stimulated a more detailed topological characterization of several social and technological networks. In particular, it has been recognized that many of these networks possess non-trivial degree correlations [4, 5, 6, 7]. The use of statistical physics tools has also evidentiated several surprising results concerning dynamical processes taking place on top of complex networks. In particular, the absence of the percolation [8, 9] and epidemic [10, 11, 12, 13, 14, 15] thresholds in uncorrelated scale-free (SF) networks has hit the community because of its potential practical implications. The absence of the percolative threshold, indeed, prompts to an exceptional tolerance to random damages [16]. On the other hand, the lack of any epidemic threshold makes SF networks the ideal media for the propagation of infections, bugs, or unsolicited information [10].

While the study of uncorrelated complex networks is a fundamental step in the understanding of the physical properties of many systems[17], yet correlations may drastically change the obtained results, as several recent works addressing the effect of correlations in epidemic spreading have shown [18, 19, 20, 21, 22].

Here we want to provide a review of recent results concerning the epidemic spreading in random correlated complex networks. We will consider the susceptible-infected-susceptible (SIS) and susceptible-infected-removed (SIR) models [23, 24, 25] and we will provide an analytical description that includes two vertices degree correlations in the dynamical evolution of the infection prevalence. This will allow us to relate the presence or absence of epidemic threshold to the eigenvalue spectra of certain connectivity matrices of the networks. In particular, in the case of scale-free networks it is possible to show that for the SIS model, a SF degree distribution $P(k) \sim k^{-\gamma}$ with $2 < \gamma \leq 3$ in unstructured networks with any kind of degree correlations is a sufficient condition for a null epidemic threshold in the thermodynamic limit. For the SIR model, the same sufficient condition applies if the minimum possible degree of the graph is $k_{min} \geq 2$. The SIR model with $k_{min} = 1$ has always a null threshold unless the SF behavior is originated only by minimum degree vertices. In other words, under very general conditions, the presence of two-point degree correlations does not alter the extreme weakness of SF networks to epidemic diffusion. The present results are derived from the divergence of the nearest neighbors average degree [4], which stems from the degree detailed balance condition [21], to be satisfied in all physical networks.

7.2 Correlated Complex Networks

In the following we shall consider unstructured undirected networks, in which all vertices within a given degree class can be considered statistically equivalent. Thus our results will not apply to structured networks in which a distance or time ordering can be defined; for instance, when the small-world property is not present [26, 27]. We will consider in particular the subset of undirected *Markovian* random networks [21], that are completely defined by the degree distribution $P(k)$ ant the conditional probability $P(k' \,|\, k)$ that a vertex of degree k is connected to a vertex of degree k'. These two functions can have any form and are assumed to be normalized, i.e.

$$\sum_k P(k) = \sum_{k'} P(k' \,|\, k) = 1. \tag{7.1}$$

The term "Markovian" refers to the fact that, in our approximation, all higher-order correlation functions can be obtained as a combination of the two fundamental functions $P(k)$ and $P(k' \,|\, k)$. In fact, this approximation represents a natural step towards a more complex description of real networks. In this sense, the Erdös-Rényi (ER) model [28] (defined starting from a set of N vertices that are connected in pairs with an independent probability p) can be viewed as the zero-th order approximation, where the average degree is the only fixed parameter. The ER model is thus defined by the ensemble of all possible networks with a given average degree, but completely random at all other respects. The first-order approximation has been recently introduced by realizing that many real-world networked systems possess a more complex degree distribution than that predicted by the ER model (a Poisson distribution [29]). In this approximation the whole degree distribution, $P(k)$, is chosen as a constrain whereas the rest of properties are left at random [30, 31, 32, 17]. Even though this approximation represents a quantitative step forward, it only takes into account local properties and, therefore, it neglects possible correlations among different vertices, correlations that, on the other hand, are present in real networks [4, 7]. Thus, it is quite natural to introduce the second-order approximation as that with a fixed degree distribution, $P(k)$, and a two-point correlation function, $P(k' \,|\, k)$, but totally random to all other respects. As we will see, in this case the approximation must be carefully made since, due to the two-point correlation constrain, the fundamental functions $P(k)$ and $P(k' \,|\, k)$ must satisfy a peculiar detailed balance condition.

The degree distribution usually identifies two kinds of networks. A first class, which includes classical models of random graphs [28], is characterized by an exponentially bounded degree distribution. A second one refers to SF networks in which the degree distribution takes the form $P(k) \sim Ck^{-\gamma}$, usually with $2 < \gamma \leq 3$ [1, 2]. In this case the network shows a very high level of degree heterogeneity, signalled by unbounded degree fluctuations. Indeed, the second moment of the degree distribution, $\langle k^2 \rangle$, diverges in the thermodynamic limit $k_c \to \infty$, where k_c is the maximum degree of the network. It is worth recalling that, in growing networks, k_c is related to the network size N as $k_c \sim N^{1/(\gamma-1)}$ [2]. Noticeably, it is the large degree heterogeneity of SF networks that is at the origin of their extreme weakness towards epidemic spreading.

7.2.1 Assortative and Disassortative Mixing

A direct study of the conditional probability $P(k' \,|\, k)$ in data from real networks usually yields results that are very noisy and difficult to interpret. In order to characterize the degree correlations, it is more useful to work with the average nearest neighbors degree (ANND) of the vertices of degree k [4], defined by

$$\bar{k}_{nn}(k) \equiv \sum_{k'} k' P(k' \,|\, k), \qquad (7.2)$$

and to plot it as a function of the degree k. When two-point correlations are not present in the network, the conditional probability takes the form $P^{nc}(k' \,|\, k) = k' P(k')/ \langle k \rangle$, and the ANND reads $\bar{k}_{nn}^{nc}(k) = \langle k^2 \rangle / \langle k \rangle$, which is independent on k. On the contrary, an explicit dependence of $\bar{k}_{nn}(k)$ on k necessary implies the existence of non-trivial correlations, as often measured in real networks [4, 7]. For instance, in many social networks it is observed that vertices with high degree connect more preferably to highly connected vertices; a property referred to as "assortative mixing". In this case, $\bar{k}_{nn}(k)$ is an increasing function of k. On the opposite side, many technological and biological networks show "disassortative mixing"; i.e. highly connected vertices are preferably connected to vertices with low degree and, consequently, $\bar{k}_{nn}(k)$ is a decreasing function of k. Then, the ANND provides an easy and powerful way to quantify two-point degree correlations, avoiding the fine details contained in the full conditional probability $P(k' \,|\, k)$.

7.2.2 Degree Detailed Balance Condition

A key relation holding for all physical networks is that all edges must point from one vertex to another. This rather obvious observation turns out to have important implications since it forces the fundamental functions $P(k)$ and $P(k' \,|\, k)$ to satisfy the following degree detailed balance condition [21]

$$kP(k' \,|\, k)P(k) = k'P(k \,|\, k')P(k'). \tag{7.3}$$

This condition states that the total number of edges pointing from vertices with degree k to vertices of degree k' must be equal to the total number of edges that point from vertices with degree k' to vertices of degree k. This relation is extremely important since it constraints the possible form of the conditional probability $P(k' \,|\, k)$ once $P(k)$ is given. It may be surprising that such a detailed balance condition exists since, in fact, networks are the result of a multiplicative random process and, in principle, detailed balance conditions only holds for systems driven by additive noise [33]. In fact, the usual detailed balance condition is the same as (7.3) without the prefactors k and k'. These very prefactors account for the multiplicative character of the network and (7.3) can be viewed as a closure condition that guarantees the existence of the network. There is a simple way to derive this condition. Let N_k be the number of vertices with degree k. Obviously, $\sum_k N_k = N$ and, consequently, we can define the degree distribution as

$$P(k) = \frac{N_k}{N}. \tag{7.4}$$

The function $P(k)$ alone does not define completely the topology of the network, because it says nothing about how vertices are connected to each other. Thus, we need to define additionally the *matrix of connections* among vertices of different degrees. Let $N_{k,k'}$ be a symmetric matrix measuring the total number of edges between vertices of degree k and vertices of degree k', when $k \neq k'$, and two times the number of self-connections, when $k = k'$. It is not difficult to realize that this matrix fulfills the identities

$$\sum_{k'} N_{k,k'} = kN_k. \tag{7.5}$$

$$\sum_{k}\sum_{k'} N_{k,k'} = \langle k \rangle N. \tag{7.6}$$

The first of this relations simply states that the number of edges emanating from all vertices of degree k is kN_k, while the second indicates that the sum of all the vertices's degrees is equal to two times the number of edges. The identity (7.6) allows us to define the joint probability

$$P(k, k') = \frac{N_{k,k'}}{\langle k \rangle N}, \tag{7.7}$$

where the symmetric function $(2 - \delta_{k,k'})P(k, k')$ is the probability that a randomly chosen edge connects two vertices of degrees k and k'. The correlation coefficient computed from this joint probability has been recently used in [7] in order to quantify two-point degree correlations. The transition probability $P(k' \mid k)$, defined as the probability that an edge from a k vertex points to a k' vertex, can be easily written as

$$P(k' \mid k) = \frac{N_{k',k}}{kN_k} \equiv \frac{\langle k \rangle \, P(k, k')}{kP(k)}, \tag{7.8}$$

from where the detailed balance condition arises as a consequence of the symmetry of $P(k, k')$ (or $N_{k,k'}$).

From the degree detailed balance condition it is possible to derive some general exact results concerning the behavior of $\bar{k}_{nn}(k, k_c)$ and of $\langle \bar{k}_{nn} \rangle_k = \sum_k P(k)\bar{k}_{nn}(k, k_c)$ in SF networks [22]. In these two functions we have now made explicit the k_c dependence originated by the upper cut-off of the k sum and that must be taken into account since it is a possible source of divergences in the thermodynamic limit. The results that we will derive will turn out to be fundamental in determining the epidemic spreading behavior in these networks. Let us start by multiplying by a k factor both terms of (7.3) and summing over k' and k. We obtain

$$\langle k^2 \rangle = \sum_{k'} k'P(k') \sum_{k} kP(k \mid k'), \tag{7.9}$$

In SF networks with $2 < \gamma < 3$ we have that the second moment of the degree distribution diverges as $\langle k^2 \rangle \sim k_c^{3-\gamma}$ [4]. We thus obtain from (7.9), using the definition (7.2),

$$\sum_{k'} k'P(k')\bar{k}_{nn}(k', k_c) \simeq \frac{C}{(3 - \gamma)}k_c^{3-\gamma}, \tag{7.10}$$

---

[4] For $\gamma = 3$ the second moment diverges as $\langle k^2 \rangle \sim \ln k_c$ but the argument, though more involved, is still valid.

where C is the constant prefactor from the degree distribution. In the case of disassortative mixing [7], the function $\bar{k}_{nn}(k', k_c)$ is decreasing with k' and, since $k'P(k')$ is an integrable function, the l.h.s. of (7.10) has no divergence related to the sum over k'. This implies that the divergence must be contained in the k_c dependence of $\bar{k}_{nn}(k', k_c)$. In other words, the function $\bar{k}_{nn}(k', k_c) \to \infty$ for $k_c \to \infty$ in a non-zero measure set. In the case of assortative mixing, $\bar{k}_{nn}(k', k_c)$ is an increasing function of k' and, depending on its rate of growth, there may be singularities associated to the sum over k'. Therefore, this case has to be analyzed in detail. Let us assume that the ANND grows as $\bar{k}_{nn}(k', k_c) \simeq \alpha k'^\beta$, $\beta > 0$, when $k' \to \infty$. If $\beta < \gamma - 2$, again there is no singularity related to the sum over k' and the previous argument for disassortative mixing holds. When $\gamma - 2 \leq \beta < 1$ there is a singularity coming from the sum over k' of the type $\alpha k_c^{\beta - (\gamma - 2)}$. However, since (7.10) comes from an identity, the singularity on the l.h.s. must match both the exponent of k_c and the prefactor on the r.h.s. In the case $\gamma - 2 \leq \beta < 1$, the singularity coming from the sum is not strong enough to match the r.h.s. of (7.10) since $\beta - (\gamma - 2) < 3 - \gamma$. Thus, the function $\bar{k}_{nn}(k', k_c)$ must also diverge when $k_c \to \infty$ in a non-zero measure set. Finally, when $\beta > 1$ the singularity associated to the sum is too strong, forcing the prefactor to scale as $\alpha \simeq r k_c^{1-\beta}$ and the ANND as $\bar{k}_{nn}(k', k_c) \simeq r k_c^{1-\beta} k'^\beta$. It is easy to realize that $r \leq 1$, since the ANND cannot be larger than k_c. Plugging the $\bar{k}_{nn}(k', k_c)$ dependence into (7.10) and simplifying common factors, we obtain the identity at the level of prefactors

$$\frac{r}{2 - \gamma + \beta} = \frac{1}{3 - \gamma}. \tag{7.11}$$

Since $\beta > 1$ and $r < 1$, the prefactor in the l.h.s. of (7.11) is smaller than the one of the r.h.s. This fact implies that the tail of the distribution in the l.h.s. of (7.10) cannot account for the whole divergence of its r.h.s. This means that the sum is not the only source of divergences and, therefore, the ANND must diverge at some other point.

In summary, the function $\bar{k}_{nn}(k', k_c)$ must diverge when $k_c \to \infty$ in a non-zero measure set independently of the correlation behavior. The large k_c singularity of the ANND can then be used to evaluate the quantity

$$\langle \bar{k}_{nn} \rangle_N = \sum_k P(k) \bar{k}_{nn}(k, k_c), \tag{7.12}$$

where we have explicitly considered k_c as a growing function of the network size N. The r.h.s. of this equation is a sum of positive terms and diverges with k_c at least as $\bar{k}_{nn}(k, k_c)$ both in the disassortative or assortative cases. In other words, *all SF networks with $2 < \gamma \leq 3$ must present a $\langle \bar{k}_{nn} \rangle_N \to \infty$ for $N \to \infty$.* This statement is independent of the structure of the correlations present in the network. The quantity $\langle \bar{k}_{nn} \rangle_N$ is particularly useful in model analysis and real data measurements. Degree correlation functions can be measured in several networks, but measurements are always performed in the presence of a finite k_c

that does not allow to exploit the singularity of the function $\bar{k}_{nn}(k, k_c)$. The most convenient way to exploit the infinite size singularity is therefore to measure the $\langle \bar{k}_{nn} \rangle_N$ for increasing network sizes.

The quantity $\langle \bar{k}_{nn} \rangle_N$ is very important in defining the properties of spreading processes in networks since it measures the number of individuals that can be infected in a few contagions. We shall discuss this point in relation to some specific epidemic models in the next sections.

7.3 The SIS Model

As a first prototypical example for examining the properties of epidemic dynamics in SF networks we consider the susceptible-infected-susceptible (SIS) model [23], in which each vertex represents an individual of the population and the edges represent the physical interactions among which the infection propagates. Each individual can be either in a susceptible or infected state. Susceptible individuals become infected with probability λ if at least one of the neighbors is infected. Infected vertices, on the other hand, recover and become susceptible again with probability one. A different recovery probability can be considered by a proper rescaling of λ and the time. This model is conceived for representing endemic infections which do not confer permanent immunity, allowing individuals to go through the stochastic cycle susceptible $\rightarrow$ infected $\rightarrow$ susceptible by contracting the infection over and over again.

7.3.1 Uncorrelated Homogeneous Networks

In uncorrelated homogeneous networks, in which each vertex has more or less the same number of connections, $k \simeq \langle k \rangle$, a general result states the existence of a finite epidemic threshold, separating an infected (endemic) phase, with a finite average density of infected individuals, from a healthy phase, in which the infection dies out exponentially fast [25]. This is for instance the case of random networks with exponentially bounded degree distribution.

This result can be recovered by considering the dynamical evolution of the average density of infected individuals $\rho(t)$ (the prevalence) present in the network. The SIS model in homogeneous uncorrelated networks at a mean-field level is described by the following rate equation [11]

$$\frac{d\rho(t)}{dt} = -\rho(t) + \lambda \langle k \rangle \rho(t) \left[1 - \rho(t)\right]. \qquad (7.13)$$

In this equation we have neglected higher order terms, since we are interested in the onset of the endemic state, close to the point $\rho(t) \sim 0$. Also, we have neglected correlations among vertices. That is, the probability of infection of a new vertex—the second term in (7.13)—is proportional to the infection rate λ, to the probability that a vertex is healthy, $1 - \rho(t)$, and to the probability that a edge in a healthy vertex points to an infected vertex. This last quantity, assuming the *homogeneous mixing hypothesis*[5], is approximated for homogeneous networks as $\langle k \rangle \, \rho(t)$, i.e. proportional to the average number of connections and to the density of infected individuals. From (7.13) it can be proved the existence of an epidemic threshold $\lambda_c = \langle k \rangle^{-1}$ [25], such that $\rho = 0$ if $\lambda < \lambda_c$, while $\rho \sim (\lambda - \lambda_c)$ if $\lambda \geq \lambda_c$. In this context, it is easy to recognize that the SIS model is a generalization of the *contact process* model, widely studied as the paradigmatic example of an absorbing-state phase transition to a unique absorbing state [34].

7.3.2 Uncorrelated Complex Networks

For general complex networks, in which large degree fluctuations and correlations might be allowed, we must relax the homogeneous hypothesis made in writing (7.13) and work instead with the relative density $\rho_k(t)$ of infected vertices with given degree k; i.e. the probability that a vertex with k edges is infected. Following [10, 11], the rate equation for $\rho_k(t)$ can be written as

$$\frac{d\rho_k(t)}{dt} = -\rho_k(t) + \lambda k \left[1 - \rho_k(t)\right] \Theta_k(t). \tag{7.14}$$

In this case, the creation term is proportional to the spreading rate λ, the density of healthy sites $1 - \rho_k(t)$, the degree k, and the variable $\Theta_k(t)$, that stands for the probability that an edge emanating from a vertex of degree k points to an infected site. In the case of an uncorrelated random network, considered in [10, 11], the probability that a edge points to a vertex with k connections is equal to $kP(k)/\langle k \rangle$ [17]. This yields a $\Theta_k = \Theta^{\text{nc}}$ independent of k that reads as

$$\Theta^{\text{nc}} = \frac{1}{\langle k \rangle} \sum_{k'} k'P(k')\rho_{k'}(t). \tag{7.15}$$

Substituting the expression (7.15) into (7.14), it is possible to find the steady state solution where Θ^{nc} is now a function of λ alone [10, 11] by the following self-consistent equation:

$$\Theta^{\text{nc}} = \frac{1}{\langle k \rangle} \sum_{k} kP(k) \frac{\lambda k \Theta^{\text{nc}}}{1 + \lambda k \Theta^{\text{nc}}}. \tag{7.16}$$

A non-zero stationary prevalence ($\rho_k \neq 0$) is obtained when the r.h.s. and the l.h.s. of (7.16), expressed as a function of Θ^{nc}, cross in the interval $0 < \Theta^{\text{nc}} \leq 1$, allowing a nontrivial solution. It is easy to realize that this corresponds to the inequality

---

[5] The homogeneous mixing hypothesis [23] states that the force of the infection (the *per capita* rate of acquisition of the disease by the susceptible individuals) is proportional to the density of infected individuals.

$$\frac{d}{d\Theta^{nc}}\left(\frac{1}{\langle k \rangle}\sum_k kP(k)\frac{\lambda k\Theta^{nc}}{1+\lambda k\Theta^{nc}}\right)\Bigg|_{\Theta^{nc}=0} \geq 1 \qquad (7.17)$$

being satisfied. The value of λ yielding the equality in (7.17) defines the critical epidemic threshold λ_c, that is given for uncorrelated random networks by

$$\lambda_c^{nc} = \frac{\langle k \rangle}{\langle k^2 \rangle}. \qquad (7.18)$$

In uncorrelated and infinite SF networks with $\gamma \leq 3$, we therefore have $\langle k^2 \rangle = \infty$, and correspondingly $\lambda_c^{nc} = 0$. This is a very relevant result, signalling that the high heterogeneity of SF networks makes them extremely weak with respect to infections. These results have several implications in human and computer virus epidemiology [35].

7.3.3 Correlated Complex Networks

For a general network in which the degrees of the vertices are correlated, the above formalism is not correct, since we are not considering the effect of the degree k into the expression for Θ_k. This effect can be taken into account, however, for Markovian networks, whose correlations are completely defined by the conditional probability $P(k' \mid k)$. In this case, it is easy to realize that the correct factor Θ_k can be written as

$$\Theta_k(t) = \sum_{k'} P(k' \mid k)\rho_{k'}(t), \qquad (7.19)$$

that is, the probability that an edge in a vertex of degree k is pointing to an infected vertex is proportional to the probability that any edge points to a vertex with degree k', times the probability that this vertex is infected, $\rho_{k'}(t)$, averaged over all the vertices connected to the original vertex. Equations (7.14) and (7.19) define together the mean-field equation describing the SIS model on Markovian complex networks,

$$\frac{d\rho_k(t)}{dt} = -\rho_k(t) + \lambda k\left[1 - \rho_k(t)\right]\sum_{k'} P(k' \mid k)\rho_{k'}(t). \qquad (7.20)$$

It must be stressed that this equation is valid only if the network has no structure; i.e. the only relevant variable is the degree k. This implies that all vertices within a given degree class are statistically equivalent. This is not the case, for instance, of regular lattices or structured networks [26, 27] in which a spatial ordering is constraining the connectivity among vertices. The exact solution of (7.20) can be difficult to find, depending on the particular form of $P(k' \mid k)$. However, it is possible to extract the value of the epidemic threshold by analyzing the stability of the steady-state solutions. Of course, the healthy state $\rho_k = 0$ is one solution. For small ρ_k, we can linearize (7.20), getting

$$\frac{d\rho_k(t)}{dt} \simeq \sum_{k'} L_{kk'} \rho_{k'}(t). \qquad (7.21)$$

In the previous equation we have defined the Jacobian matrix $\mathbf{L} = \{L_{kk'}\}$ by

$$L_{kk'} = -\delta_{kk'} + \lambda k P(k' \,|\, k), \qquad (7.22)$$

where $\delta_{kk'}$ is the Kronecker delta symbol. The solution $\rho_k = 0$ will be unstable if there exists at least one positive eigenvalue of the Jacobian matrix $\mathbf{L}$. Let us consider the *connectivity matrix* $\mathbf{C}$, defined by $C_{kk'} = kP(k' \,|\, k)$. Using the symmetry condition (7.3), it is easy to check that if v_k is an eigenvector of $\mathbf{C}$, with eigenvalue Λ, then $P(k)v_k$ is an eigenvector of the transposed matrix $\mathbf{C}^T$ with the same eigenvalue. From here it follows immediately that all the eigenvalues of $\mathbf{C}$ are real. Let Λ_m be the largest eigenvalue of $\mathbf{C}$. Then, the origin will be unstable whenever $-1 + \lambda \Lambda_m > 0$, which defines the epidemic threshold

$$\lambda_c = \frac{1}{\Lambda_m}, \qquad (7.23)$$

above which the solution $\rho_k = 0$ is unstable, and another non-zero solution takes over as the actual steady-state—the endemic state.

It is instructive to see how this general formalism recovers previous results [10, 11], implicitly obtained for random uncorrelated networks. For any random network, in which there are no correlations among the degrees of the vertices, we have that the connectivity matrix is given by $C_{kk'}^{\mathrm{nc}} = kP(k'/k) \equiv kk'P(k')/\langle k \rangle$, since the probability that an edge points to a vertex of connectivity k' is proportional to $k'P(k')$. It is easy to check that the matrix $\{C_{k'k}^{\mathrm{nc}}\}$ has a unique eigenvalue $\Lambda_m^{\mathrm{nc}} = \langle k^2 \rangle / \langle k \rangle$, corresponding to the eigenvector $v_k^{\mathrm{nc}} = k$, from where we recover the now established result (7.18).

7.3.4 Correlated Scale-Free Networks

The absence of an epidemic threshold in SF uncorrelated networks is an extremely important question that prompts to a possible weakness of many real-world networks. Particularly important is for the case of digital viruses spreading on the Internet [10, 4] and sexually transmitted diseases diffusing on the web of sexual contacts [36]. Both these networks show, in fact, SF properties that would imply the possibility of major epidemic outbreaks even for infections with a very low transmission rate. Immunization policies as well must be radically changed in the case that a network has a null epidemic threshold [37, 38, 39].

In view of the relevance of this framework, it is extremely important to study to which extent the presence of correlations are altering these results. The main question is therefore which conditions on the degree correlations of SF networks preserve the lack of a critical threshold. In the case of correlated networks, we have shown that the epidemic threshold is the inverse of the largest eigenvalue of the connectivity matrix $\mathbf{C}$. The absence of an epidemic threshold thus corresponds to a divergence of the largest eigenvalue of the connectivity matrix $\mathbf{C}$ in the limit of an infinite network size $N \to \infty$. In order to provide some general statement on the conditions for such a divergence we can make use of the Frobenius theorem for non-negative irreducible matrices [40]. This theorem states the existence of the largest eigenvalue of any non-negative irreducible matrix, eigenvalue which is simple, real, positive, and has a positive eigenvector. In our case the matrix of interest is the connectivity matrix that is non-negative and irreducible. The irreducible property of the connectivity matrix is a simple consequence of the fact that all the degree classes in the network are accessible. That is, starting from the degree class k it is always possible to find a path of edges that connects this class to any other class k' of the network. If this is not the case it means that the network is built up of disconnected irreducible subnetworks and, therefore, we can apply the same line of reasoning to each subnetwork[6]. From the Frobenius theorem [40] it can be proved that the maximum eigenvalue, Λ_m, of any non-negative irreducible matrix, $A_{kk'}$, satisfies the inequality

$$\Lambda_m \geq \min_k \frac{1}{\psi(k)} \sum_{k'} A_{kk'} \psi(k'), \qquad (7.24)$$

where $\{\psi(k)\}$ is any positive vector. In particular, by setting $\mathbf{A} = \mathbf{C}^2$ and $\psi(k) = k$ we obtain the inequality

$$\Lambda_m^2 \geq \min_k \sum_{k'} \sum_{\ell} k'\ell P(\ell\,|\,k)P(k'\,|\,\ell). \qquad (7.25)$$

This inequality relates the lower bound of the largest eigenvalue Λ_m to the degree correlation function and allows to find a sufficient condition for the absence of the epidemic threshold. By noting that $\sum_{k'} k'P(k'\,|\,\ell) = \bar{k}_{nn}(\ell, k_c)$, we obtain the inequality

$$\Lambda_m^2 \geq \min_k \sum_{\ell} \ell P(\ell\,|\,k)\bar{k}_{nn}(\ell, k_c). \qquad (7.26)$$

The r.h.s. of this equation is a sum of positive terms, and by recalling the divergence of the ANND with k_c, we readily obtain that $\Lambda_m \geq \infty$ for all networks with diverging $\langle k^2 \rangle$ both in the disassortative or assortative cases[7]. The divergence of Λ_m implies on its turn that the SIS epidemic threshold vanishes, in the thermodynamic limit, in all SF networks with assortative or disassortative mixing if the degree distribution has a diverging second moment; *i.e. a SF degree*

---

[6] Notice that being irreducible is not equivalent to being fully connected at the vertex to vertex level, but at the class to class level.

[7] One may argue that, since we are calculating a minimum for k, if the transition probability $P(\ell\,|\,k_0)$ is zero at some point k_0, this minimum is zero. In this case it is possible to show that repeating the same argument with $\mathbf{C}^3$ instead of $\mathbf{C}^2$ provides us an inequality that avoids this problem, [21].

distribution with exponent $2 < \gamma \leq 3$ is a sufficient condition for the absence of an epidemic threshold for the SIS model in unstructured networks with arbitrary two-point degree correlation function.

In physical terms, the absence of the epidemic threshold is related to the divergence of $\langle \bar{k}_{nn} \rangle_N$ in SF networks. In homogeneous networks, where $\langle \bar{k}_{nn} \rangle_N \simeq \langle k \rangle$, the epidemic spreading properties can be related to the average degree. In SF networks, however, the focus shifts to the possibility of infecting a large number of individuals in a finite number of contagions. The fact that an infected vertex has a very low degree is not very important if a hub of the network that provides connectivity to a large number of vertices is a few hops away. The infection can, in this case, very easily access a very large number of individuals in a short time. In SF networks is the ANND that takes into account more properly the level of degree fluctuations and thus rules the epidemic spreading dynamics.

It is worth stressing that the divergence of $\langle \bar{k}_{nn} \rangle_N$ is ensured by the degree detailed balance condition alone, and it is a very general result holding for all SF networks with $2 < \gamma \leq 3$. On the contrary, the SF behavior with $2 < \gamma \leq 3$ is a sufficient condition for the lack of epidemic threshold only in networks with general two-point degree correlations and in absence of higher-order correlations. The reason is that the relation between the epidemic threshold and the maximum eigenvalue of the connectivity matrix only holds for these classes of networks. Higher order correlations, or the presence of an underlying metric in the network [27], can modify the rate equation at the basis of the SIS model and may invalidate the present discussion.

7.4 The SIR Model

The susceptible-infected-removed model (SIR) [23] represents the other paradigmatic example of epidemic dynamics. Unlike in the SIS model, in this case infected individuals fall, after some random time, into a removed state where they cannot neither become infected again nor infect other individuals. This model tries to mimic real epidemics where individuals, after being infected, acquire permanent immunity or, in the worst case, die.

The SIR model is defined as follows. Individuals can only exist in three different states, namely, susceptible, infected, or removed. Susceptible individuals become infected with probability λ if at least one of their neighbors is infected. On the other hand, infected individuals spontaneously fall in the removed state with probability μ, which without lack of generality we set equal to unity. The main difference between both models is that whereas in the SIS, for $\lambda > \lambda_c$, the epidemics reaches a steady state, in the SIR the epidemics always dies and reaches eventually a state with zero density of infected individuals. The epidemic prevalence is thus defined in this case as the total number of infected individuals in the whole epidemic process.

7.4.1 Uncorrelated Homogeneous Networks

In a homogeneous system, the SIR model can be described in terms of the densities of susceptible, infected, and removed individuals, $S(t)$, $\rho(t)$, and $R(t)$, respectively, as a function of time. These three quantities are linked through the normalization condition

$$S(t) + \rho(t) + R(t) = 1, \tag{7.27}$$

and they obey the following system of differential equations:

$$\frac{dS}{dt} = -\lambda \langle k \rangle \rho S,$$

$$\frac{d\rho}{dt} = -\rho + \lambda \langle k \rangle \rho S, \tag{7.28}$$

$$\frac{dR}{dt} = \rho.$$

These equations can be interpreted as follows: infected individuals decay into the removed class at a unity rate, while susceptible individuals become infected at a rate proportional to both the densities of infected and susceptible individuals. Here, λ is the microscopic spreading (infection) rate, and $\langle k \rangle$ is the number of contacts per unit time that is supposed to be constant for the whole population. In writing this last term of the equations we have assumed again, as in the case of the SIS model, the homogeneous mixing hypothesis [23],

The most significant prediction of this model is the presence of a nonzero epidemic threshold λ_c [24]. If the value of λ is above λ_c, $\lambda > \lambda_c$, the disease spreads and infects a finite fraction of the population. On the other hand, when λ is below the threshold, $\lambda < \lambda_c$, the total number of infected individuals (the epidemic prevalence), $R_\infty = \lim_{t \to \infty} R(t)$, is infinitesimally small in the limit of very large populations. In order to see this point, let us consider the set of equations (7.28). Integrating the equation for $S(t)$ with the initial conditions $R(0) = 0$ and $S(0) \simeq 1$ (i.e., assuming $\rho(0) \simeq 0$, a very small initial concentration of infected individuals), we obtain

$$S(t) = e^{-\lambda \langle k \rangle R(t)}. \tag{7.29}$$

Combining this result with the normalization condition (7.27), we observe that the total number of infected individuals R_∞ fulfills the following self-consistent equation:

$$R_\infty = 1 - e^{-\lambda \langle k \rangle R_\infty}. \tag{7.30}$$

While $R_\infty = 0$ is always a solution of this equation, in order to have a nonzero solution the following condition must be fulfilled:

$$\frac{d}{dR_\infty} \left(1 - e^{-\lambda \langle k \rangle R_\infty} \right) \bigg|_{R_\infty = 0} \geq 1. \tag{7.31}$$

This condition is equivalent to the constraint $\lambda \geq \lambda_c$, where the epidemic threshold λ_c takes the value $\lambda_c = \langle k \rangle^{-1}$. Performing a Taylor expansion at $\lambda = \lambda_c$ it is then possible to obtain the epidemic prevalence behavior $R_\infty \sim (\lambda - \lambda_c)$ (valid above the epidemic threshold). From the point of view of the physics of non-equilibrium phase transition, it is easy to recognize that the SIR model is a generalization of the *dynamical percolation* model, that has been extensively studied in the context of absorbing-state phase transitions [34].

7.4.2 Uncorrelated Complex Networks

In order to take into account the heterogeneity induced by the presence of vertices with different degree, we consider the time evolution of the magnitudes $\rho_k(t)$, $S_k(t)$, and $R_k(t)$, which are the density of infected, susceptible, and removed vertices of degree k at time t, respectively [14, 15]. These variables are connected by means of the normalization condition

$$\rho_k(t) + S_k(t) + R_k(t) = 1. \tag{7.32}$$

Global quantities such as the epidemic prevalence can be expressed as an average over the various degree classes; for example, we define the total number of removed individuals at time t by $R(t) = \sum_k P(k)R_k(t)$, and the prevalence as $R_\infty = \lim_{t \to \infty} R(t)$. At the mean-field level, for random uncorrelated networks, these densities satisfy the following set of coupled differential equations:

$$\frac{d\rho_k(t)}{dt} = -\rho_k(t) + \lambda k S_k(t) \Theta^{\mathrm{nc}}(t), \tag{7.33}$$

$$\frac{dS_k(t)}{dt} = -\lambda k S_k(t) \Theta^{\mathrm{nc}}(t), \tag{7.34}$$

$$\frac{dR_k(t)}{dt} = \rho_k(t). \tag{7.35}$$

The factor $\Theta^{\mathrm{nc}}(t)$ represents the probability that any given edge points to an infected vertex and is capable of transmitting the disease. This quantity can be computed in a self-consistent way: The probability that an edge points to an infected vertex with degree k' is proportional to $k'P(k')$. However, since the infected vertex under consideration received the disease through a particular edge that cannot be used for transmission anymore (since it points back to a previously infected individual) the correct probability must consider one less edge. Therefore,

$$\Theta^{\mathrm{nc}}(t) = \frac{1}{\langle k \rangle} \sum_k (k-1)P(k)\rho_k(t). \tag{7.36}$$

The equations (7.33), (7.34), (7.35), and (7.36), combined with the initial conditions $R_k(0) = 0$, $\rho_k(0) = \rho_k^0$, and $S_k(0) = 1 - \rho_k^0$, completely define the SIR model on any uncorrelated complex network with degree distribution $P(k)$. We

will consider in particular the case of a homogeneous initial distribution of infected individuals, $\rho_k^0 = \rho^0$. In this case, in the limit $\rho^0 \to 0$, we can substitute
$\rho_k(0) \simeq 0$ and $S_k(0) \simeq 1$. Under this approximation, (7.34) and (7.35) can be
directly integrated, yielding

$$S_k(t) = e^{-\lambda k \phi(t)}, \qquad R_k(t) = \int_0^\infty \rho_k(\tau)dt, \tag{7.37}$$

where we have defined the auxiliary function

$$\phi(t) = \int_0^t \Theta^{\mathrm{nc}}(\tau)d\tau = \frac{1}{\langle k \rangle} \sum_k (k-1)P(k)R_k(t). \tag{7.38}$$

In order to get a closed relation for the total density of infected individuals, it
results more convenient to focus on the time evolution of the averaged magnitude
$\phi(t)$. To this purpose, let us compute its time derivative:

$$\frac{d\phi(t)}{dt} = 1 - \frac{1}{\langle k \rangle} - \phi(t) - \frac{1}{\langle k \rangle} \sum_k (k-1)P(k)e^{-\lambda k \phi(t)}, \tag{7.39}$$

where we have introduced the time dependence of $S_k(t)$ obtained in (7.37). Once
solved (7.39), we can obtain the total epidemic prevalence R_∞ as a function of
$\phi_\infty = \lim_{t \to \infty} \phi(t)$. Since $R_k(\infty) = 1 - S_k(\infty)$, we have

$$R_\infty = \sum_k P(k)\left(1 - e^{-\lambda k \phi_\infty}\right). \tag{7.40}$$

For a general $P(k)$ distribution, (7.39) cannot be generally solved in a closed
form. However, we can still get useful information on the infinite time limit; i.e.
at the end of the epidemics. Since we have that $\rho_k(\infty) = 0$, and consequently
$\lim_{t \to \infty} d\phi(t)/dt = 0$, we obtain from (7.39) the following self-consistent equation
for ϕ_∞:

$$\phi_\infty = 1 - \frac{1}{\langle k \rangle} - \frac{1}{\langle k \rangle} \sum_k (k-1)P(k)e^{-\lambda k \phi_\infty}. \tag{7.41}$$

The value $\phi_\infty = 0$ is always a solution. In order to have a non-zero solution, the
condition

$$\frac{d}{d\phi_\infty}\left(1 - \frac{1}{\langle k \rangle} - \frac{1}{\langle k \rangle} \sum_k (k-1)P(k)e^{-\lambda k \phi_\infty}\right)\Bigg|_{\phi_\infty=0} \geq 1 \tag{7.42}$$

must be fulfilled. This relation implies

$$\frac{\lambda}{\langle k \rangle} \sum_k k(k-1)P(k) \geq 1, \tag{7.43}$$

which defines the epidemic threshold

$$\lambda_c^{\mathrm{nc}} = \frac{\langle k \rangle}{\langle k^2 \rangle - \langle k \rangle}, \tag{7.44}$$

below which the epidemic prevalence is null, and above which it attains a finite value. It is interesting to notice that this is precisely the same value found for the percolation threshold in generalized networks [8, 9]. This is hardly surprising since, as it is well known [41], the SIR model can be mapped to a bond percolation process.

7.4.3 Correlated Complex Networks

In order to work out the SIR model in Markovian networks it is easier to consider the rate equations for the quantities $N_k^I(t)$ and $N_k^R(t)$, defined as the number of infected and removed individuals of degree k, present at time t, respectively. From this two quantities we can easily recover the densities $\rho_k(t)$ and $R_k(t)$ as

$$\rho_k(t) = \frac{N_k^I(t)}{N_k}, \qquad R_k(t) = \frac{N_k^R(t)}{N_k}, \qquad S_k(t) = 1 - \rho_k(t) - R_k(t), \tag{7.45}$$

where $N_k = NP(k)$ is the number of vertices with degree k. The rate equations for $N_k^I(t)$ and $N_k^R(t)$ are then given by

$$\frac{dN_k^I(t)}{dt} = -N_k^I(t) + \lambda S_k(t)\Gamma_k(t), \tag{7.46}$$

$$\frac{dN_k^R(t)}{dt} = N_k^I(t), \tag{7.47}$$

where we have defined the function

$$\Gamma_k(t) \equiv \sum_{k'} N_{k'}^I(t)(k' - 1)P(k \,|\, k') \tag{7.48}$$

In this case, the creation of new infected individuals—the second term in the r.h.s. of (7.46)—is proportional to the number of infected individuals of degree k', $N_{k'}^I(t)$, the probability that a vertex of degree k is susceptible, $S_k(t)$, and the average number of edges pointing from these infected vertices to vertices of degree k, $(k' - 1)P(k \,|\, k')$, all averaged for all the vertices of degree k'. This last term takes into account that one of the edges is not available for transmitting the infection, since it was used to infect the vertex considered. Dividing (7.46) by N_k and making use of the detailed balance condition (7.3) we find the rate equations for the relative densities as

$$\frac{dR_k(t)(t)}{dt} = \rho_k(t), \tag{7.49}$$

$$\frac{d\rho_k(t)}{dt} = -\rho_k(t) + \lambda k S_k(t)\Theta_k(t), \tag{7.50}$$

where, for a Markovian network, the factor $\Theta_k(t)$ takes the form

$$\Theta_k(t) = \sum_{k'} \rho_{k'}(t) \frac{k'-1}{k'} P(k' \mid k) \tag{7.51}$$

Again, it must be stressed that no structure is allowed in the network in order for this equations to represent a valid formulation of the SIR model.

In order to extract information about the epidemic threshold, we proceed similarly to the SIS model, performing a linear stability analysis. For time $t \to 0$, that corresponds to small ρ_k, $R_k \simeq 0$ and $S_k \simeq 1$, (7.50) and (7.51) can be written as

$$\frac{d\rho_k(t)}{dt} \simeq \sum_{k'} \tilde{L}_{kk'} \rho_{k'}(t), \tag{7.52}$$

where the Jacobian matrix $\tilde{\mathbf{L}} = \{\tilde{L}_{kk'}\}$ can be written as

$$\tilde{L}_{kk'} = -\delta_{kk'} + \lambda \frac{k(k'-1)}{k'} P(k' \mid k). \tag{7.53}$$

In order to infect a finite fraction of individuals, we need the solution $\rho_k = 0$ to be unstable, which happens if there is at least one positive eigenvalue of the Jacobian matrix $\tilde{\mathbf{L}}$. Defining the matrix $\tilde{\mathbf{C}} = \{\tilde{C}_{kk'}\}$, with elements

$$\tilde{C}_{kk'} = \frac{k(k'-1)}{k'} P(k' \mid k), \tag{7.54}$$

we know from the Frobenius theorem, since it is positive and provided that it is reducible at the degree class level, that it has a largest eigenvalue $\tilde{\Lambda}_m$ that is real and positive. Thus, the solution $\rho_k = 0$ of (7.52) is stable whenever $-1 + \lambda \tilde{\Lambda}_m < 0$. This relation defines the epidemic threshold for the SIR model in Markovian networks

$$\lambda = \frac{1}{\tilde{\Lambda}_m}. \tag{7.55}$$

In the case of a random uncorrelated network, we have that $\tilde{C}_{kk'}^{\mathrm{nc}} = k(k'-1)P(k')/\langle k \rangle$. It can be easily seen that this matrix has a unique eigenvalue $\tilde{\Lambda}_m^{\mathrm{nc}} = \langle k^2 \rangle / \langle k \rangle - 1$, corresponding to the eigenvalue $\tilde{v}_k^{\mathrm{nc}} = k$, thus recovering the previous result (7.44) obtained for this kind of networks.

The relation between the SIR model and percolation in correlated complex networks can be closed by noticing that the relevant parameter in this last problem is the largest eigenvalue of the matrix $\mathbf{C}^{\mathrm{perc}} = \{C_{kk'}^{\mathrm{perc}}\}$, with elements $C_{kk'}^{\mathrm{perc}} = (k'-1)P(k' \mid k)$, as pointed out in [42]. It is easy to check that if v_k^{perc} is an eigenvector of $\mathbf{C}^{\mathrm{perc}}$ with eigenvalue Λ, then $\tilde{v}_k = k \, v_k^{\mathrm{perc}}$ is an eigenvector of $\tilde{\mathbf{C}}$ with the same eigenvalue. Then, the eigenvalues of $\mathbf{C}^{\mathrm{perc}}$ and $\tilde{\mathbf{C}}$ coincide, yielding in this way the same description and an identical critical point.

7.4.4 Correlated Scale-Free Networks

The discussion of the absence of epidemic threshold of the SIS in SF networks with any sort of degree correlations can be easily extended to the SIR model, taking again advantage of the Frobenius theorem. In this case, in the general inequality given by (7.24), we set $\mathbf{A} = \tilde{\mathbf{C}}^2$ and $\psi(k) = k$, obtaining

$$\tilde{\Lambda}_m^2 \geq \min_k \sum_{k'} \sum_{\ell} (\ell - 1) P(\ell \,|\, k)(k' - 1) P(k' \,|\, \ell). \qquad (7.56)$$

Given that $\sum_{k'} (k' - 1) P(k' \,|\, \ell) = \bar{k}_{nn}(\ell, k_c) - 1$, the previous inequality reads

$$\tilde{\Lambda}_m^2 \geq \min_k \sum_{\ell} (\ell - 1) P(\ell \,|\, k) \left[\bar{k}_{nn}(\ell, k_c) - 1 \right]. \qquad (7.57)$$

As in the case of the SIS model, the divergence of the ANND with k_c in the thermodynamic limit, ensures the divergence of the eigenvalue $\tilde{\Lambda}_m$. Therefore, a SF degree distribution with diverging second moment is a sufficient condition for the absence of an epidemic threshold also for the SIR model if the minimum degree of the network is $k_{min} \geq 2$. The only instance in which we can have an infinite $\bar{k}_{nn}(\ell, k_c)$ with a finite eigenvalue $\tilde{\Lambda}_m$ is when the divergence of $\bar{k}_{nn}$ is accumulated in the degree $k = 1$ and results canceled by the term $\ell - 1$ in (7.57). This situation happens when the SF behavior of the degree distribution is just due to vertices with a single edge that form star-like structures by connecting on a few central vertices. Explicit examples of this situation are provided in [27, 42].

7.5 Conclusions

In this paper we have reviewed the analytical treatment of the epidemic SIS and SIR models in complex networks at different levels of approximation, corresponding to the different levels in which degree correlations can be taken into account. At the zero-th level, in which all the vertices are assumed to have the same degree (homogeneous networks), we observe the presence of an epidemic threshold, separating an active or endemic phase from an inactive or healthy phase, that is inversely proportional to the average degree $\langle k \rangle$. At this level of approximation, both models render the same result, thus showing a high degree of universality. At the first order approximation level, in which vertices are allowed to have a different degree, drawn from a specified degree distribution $P(k)$, but are otherwise random, we obtain epidemic thresholds that are inversely proportional to the degree fluctuations $\langle k^2 \rangle$. The remarkable fact about this result is that the epidemic threshold vanishes for SF networks with characteristic exponent $2 < \gamma \leq 3$ in the limit of an infinitely large network. Finally, in the second order approximation level, in which degree correlations are explicitly controlled by the conditional probability $P(k' \,|\, k)$ that a vertex of degree k is connected to a vertex of degree k', our analysis yields that the epidemic threshold in the SIS and SIR models is inversely proportional to the largest eigenvalue of the connectivity matrices $C_{kk'} = kP(k' \mid k)$ and $\tilde{C}_{kk'} = k(k'-1)P(k' \mid k)/k'$, respectively. In the case of the SIR model we recover the mapping with percolation at the level of correlations exclusively among nearest neighbor vertices. The analysis of the divergence of the average nearest neighbors degree $\bar{k}_{nn}(k, k_c)$ with the degree cut-off k_c allows us to establish the general result that any SF degree distribution with diverging second moment is a sufficient condition for the vanishing of the epidemic threshold in the SIS model. The same sufficient condition holds in the SIR model with $k_{min} \geq 2$. The SIR model with $k_{min} = 1$ always shows the absence of an epidemic threshold with the exception of the peculiar case in which the divergence of the average nearest neighbor degree is accumulated only on the nodes of minimum degree. These results have extremely important consequences, since they imply that correlations are not able to stop an epidemic outbreak in SF networks, in opposition to previous claims, and indicates that a reduction of epidemic incidence can only be obtained by means of carefully crafted immunization strategies [37, 38, 39], or trivially through finite size effects [43].

Acknowledgments

This work has been partially supported by the European commission FET Open project COSIN IST-2001-33555. R.P.-S. acknowledges financial support from the Ministerio de Ciencia y Tecnología (Spain).

References

1. R. Albert and A.-L. Barabási. Statistical mechanics of complex networks. *Rev. Mod. Phys.* **74**, 47–97 (2002)
2. S.N. Dorogovtsev and J.F.F. Mendes. Evolution of networks. *Adv. Phys.* **51**, 1079–1187 (2002)
3. D.J. Watts and S.H. Strogatz. Collective dynamics of 'small-world' networks. *Nature* **393**, 440–442 (1998)
4. R. Pastor-Satorras, A. Vázquez, and A. Vespignani. Dynamical and correlation properties of the Internet. *Phys. Rev. Lett.* **87**, 258701 (2001)
5. A. Vázquez, R. Pastor-Satorras, and A. Vespignani. Large-scale topological and dynamical properties of Internet. *Phys. Rev. E* **65**, 066130 (2002)
6. S. Maslov and K Sneppen. Specificity and stability in topology of protein networks. *Science* **296**, 910–913 (2002)
7. M.E.J. Newman. Assortative mixing in networks. *Phys. Rev. Lett.* **89**, 208701 (2002)
8. D.S. Callaway, M.E.J. Newman, S.H. Strogatz, and D.J. Watts. Network robustness and fragility: percolation on random graphs. *Phys. Rev. Lett.* **85**, 5468–5471 (2000)
9. R. Cohen, K. Erez, D. ben Avraham, and S. Havlin. Resilience of the Internet to random breakdowns. *Phys. Rev. Lett.* **85**, 4626 (2000)

10. R. Pastor-Satorras and A. Vespignani. Epidemic spreading in scale-free networks. *Phys. Rev. Lett.* **86**, 3200–3203 (2001)
11. R. Pastor-Satorras and A. Vespignani. Epidemic dynamics and endemic states in complex networks. *Phys. Rev. E* **63**, 066117 (2001)
12. R.M. May and A.L. Lloyd. Infection dynamics on scale-free networks. *Phys. Rev. E* **64**, 066112 (2001)
13. R. Pastor-Satorras and A. Vespignani. Epidemics and immunization in scale-free networks. In *Handbook of Graphs and Networks: From the Genome to the Internet*, S. Bornholdt and H.G. Schuster, editors, 113–132. Wiley-VCH, Berlin (2002)
14. Y. Moreno, R. Pastor-Satorras, and A. Vespignani. Epidemic outbreaks in complex heterogeneous networks. R *Eur. Phys. J. B* **26**, 521–529 (2002)
15. M.E.J. Newman. Spread of epidemic diseases on networks. *Phys. Rev. E* **64**, 016128 (2002)
16. R.A. Albert, H. Jeong, and A.-L. Barabási. Error and attack tolerance of complex networks. *Nature* **406**, 378–382 (2000)
17. M.E.J. Newman. Random graphs as models of networks. In *Handbook of Graphs and Networks: From the Genome to the Internet*, S. Bornholdt and H.G. Schuster, editors. Wiley-VCH, Berlin (2002)
18. V.M. Eguíluz and K. Klemm. Epidemic threshold in structured scale-free networks. *Phys. Rev. Lett.* **89**, 108701 (2002)
19. C.P. Warren, L.M. Sander, and I.M. Sokolov. Geography in a scale-free network model. *Phys. Rev. E* **66**, 056105 (2002)
20. Ph. Blanchard, Ch.-H. Chang, and T. Krueger. Epidemic thresholds on scale-free graphs: the interplay between exponent and preferential choice, (2002). e-print cond-mat/0207319
21. M. Boguñá and R. Pastor-Satorras. Epidemic spreading in correlated complex networks. *Phys. Rev. E* **66**, 047104 (2002)
22. M. Boguñá, R. Pastor-Satorras, and A. Vespignani. Absence of epidemic threshold in scale-free networks with degree correlations, (2002). e-print cond-mat/0208163
23. R.M. Anderson and R.M. May, *Infectious diseases in humans*, (Oxford University Press, Oxford, 1992)
24. J.D. Murray, *Mathematical biology*, (Springer Verlag, Berlin, 1993). 2nd edition
25. O. Diekmann and J.A.P Heesterbeek, *Mathematical epidemiology of infectious diseases: model building, analysis and interpretation*, (John Wiley & Sons, New York, 2000)
26. K. Klemm and V.M. Eguíluz. Highly clustered scale-free networks. *Phys. Rev. E* **65**, 036123 (2002)
27. Y. Moreno and A.Vázquez. Disease spreading in structured scale-free networks, (2002). e-print cond-mat/0210362
28. P. Erdös and P.Rényi. On random graphs. *Publicationes Mathematicae* **6**, 290–297 (1959)
29. B. Bollobás. Degree sequences of random graphs. *Discrete Math.* **33**, 1–19 (1981)
30. E.A. Bender and E.R. Canfield. The asymptotic number of labeled graphs with given degree distribution. *Journal of Combinatorial Theory A* **24**, 296–307 (1978)
31. M. Molloy and B. Reed. A critical point for random graphs with a given degree sequence. *Random Struct. Algorithms* **6**, 161 (1995)
32. M. Molloy and B. Reed. The size of the giant component of a random graph with a given degree sequence. *Combinatorics, Probab. Comput.* **7**, 295 (1998)
33. N.G. van Kampen, *Stochastic processes in chemistry and physics*, (North Holland, Amsterdam, 1981)

34. J. Marro and R. Dickman, *Nonequilibrium phase transitions in lattice models*, (Cambridge University Press, Cambridge, 1999)
35. A.L. Lloyd and R.M. May. How viruses spread among computers and people. *Science* **292**, 1316–1317 (2001)
36. F. Liljeros, C.R. Edling, L.A.N. Amaral, H.E. Stanley, and Y. Aberg. The web of human sexual contacts. *Nature* **411**, 907–908 (2001)
37. R. Pastor-Satorras and A. Vespignani. Immunization of complex networks. *Phys. Rev. E* **65**, 036104 (2001)
38. Z. Dezsö and A.-L. Barabási. Halting viruses in scale-free networks. *Phys. Rev. E* **65**, 055103(R) (2002)
39. R. Cohen, D. ben-Avraham, and S. Havlin. Efficient immunization of populations and computers, (2002). e-print cond-mat/0209586
40. F.R. Gantmacher, *The theory of matrices*, volume II, (Chelsea Publishing Company, New York, 1974)
41. P. Grassberger. On the critical behavior of the general epidemic process and dynamical percolation. *Math. Biosci.* **63**, 157–172 (1983)
42. A. Vázquez and Y. Moreno. Resilience to damage of graphs with degree correlations, (2002)
43. R. Pastor-Satorras and A. Vespignani. Epidemic dynamics in finite size scale-free networks. *Phys. Rev. E* **65**, 035108 (2002)

8 Food Web Structure and the Evolution of Complex Networks

Guido Caldarelli[1], Diego Garlaschelli[1,2], and Luciano Pietronero[1,3]

[1] INFM and Dipartimento di Fisica, Università "La Sapienza", P.le Aldo Moro 5, 00185 Roma, Italy
[2] INFM and Dipartimento di Fisica, Università di Siena, Banchi di Sotto 55, 53100 Siena, Italy
[3] CNR, Istituto di Acustica "O.M. Corbino", V. Fosso del Cavaliere 100, 00133 Roma, Italy

Abstract. In addition to traditional properties such as the degree distribution $P(k)$, in this work we propose two other useful quantities that can help in characterizing the topology of food webs quantitatively, namely the allometric scaling relations $C(A)$ and the branch size distribution $P(A)$ which are defined on the spanning tree of the webs. These quantities, whose use has proved relevant in characterizing other different networks appearing in nature (such as river basins, Internet, and vascular systems), are related (in the context of food webs) to the efficiency in the resource transfer and to the stability against species removal. We present the analysis of the data for both real food webs and numerical simulations of a growing network model. Our results allow us to conclude that real food webs display a high degree of both efficiency and stability due to the evolving character of their topology.

8.1 Introduction

Food webs [1, 2, 3] are an important example of complex networks [4] describing the predation interactions among species in a given environment. A food web can be defined as a *directed graph* [4, 5], which is a set of S *vertices* (each labelled by an integer number i) representing biological species and L *directed links* pointing from prey to predators. Conventionally, a set of species sharing the same predators and the same prey is merged in one *trophic species* (this is referred to as the *aggregation* of a food web, and is commonly performed in order to reduce systematic biases [6, 7]).

The exploration of food web structure is one of the major issues of modern ecology [1, 2, 3]. Understanding how communities are assembled and evolve would give a deep insight into the organization of natural ecosystems. In particular, one of the main focuses of food web theory is understanding how (or whether) topological properties of food webs change with the scale of the system (the number of species). A variety of ecological quantities are traditionally introduced in order to describe and compare different food webs. Examples are given by the fractions T, B, I of *Top*, *Basal* and *Intermediate* species, which are defined as the species with respectively no predators, no prey, and both predators and prey. These quantities also give the *prey–predator ratio*, defined as $(B + I)/(I + T)$.

Another quantity of interest is the ratio of observed to possible links. If self (cannibalistic) loops and reciprocal connections between two species are not included, this ratio is expressed by the *connectance* $c = 2L/(S^2 - S) \simeq 2L/S^2$. Otherwise, the *directed connectance* [6] $c_d = L/S^2 \simeq c/2$ is used. The scale dependence of these quantities has been investigated [2, 6, 8, 9, 10], and none of them shows a definite trend as the number of species varies. In particular, the directed connectance varies from $c_d = 0.03$ to $c_d = 0.3$ in real webs [11], deviating from the *constant connectance* [12] hypothesis which predicts c_d to be approximately constant about the value 0.1.

On the other hand, the recent exploration of network structures [4, 5] showed that several real-world networks, ranging from Internet and WWW to social and biological systems, display unexpectedly similar properties. The quantities which have been introduced to characterise network topology derive from graph theory [5]. The number of incoming and outgoing links of a vertex is called the *in-degree* k_{in} and the *out-degree* k_{out} of the vertex (their sum gives the total degree k). In the ecological context, they have a direct interpretation in terms of the number of prey and predators of a species respectively. The *degree distribution* $P(k)$ gives the probability that a randomly chosen vertex has total degree k. Also, one can introduce the *average distance* D (defined as the mean number of links required to connect two randomly chosen vertices in the network), and the *clustering coefficient* C_c (which is the probability of finding a link connecting two neighbours of a randomly chosen vertex). While random graph models [13] are characterised by a Poisson-like degree distibution and a small value of the clustering coefficient, most real networks display a *scale-free* [5] degree distribution of the form $P(k) \propto k^{-\gamma}$ and a *small-world* [14] character. The latter is defined as the simultaneous occurrence of a small average distance D and a high value of the clustering coefficient C_c. The deviation from random graphs means that real networks are rather *complex* structures shaped by non-random processes. This important aspect motivates the search for non-trivial models that can reproduce the complex topology of real systems.

Recently, the investigation of such network properties has been extended to food webs [7, 11, 15, 16, 17]. In this work, we report the results of the analysis of four empirical networks [6, 8, 9, 10], three of which are the largest ones in the ecological literature. As other authors [11, 15, 17] have independently observed, we show that (differently from other networks) the aforementioned quantities display an ambiguous behaviour in food webs. We suggest that these difficulties can be overcome by analysing food webs in a different framework, namely that of transportation networks [18, 19]. This rather natural choice provides us with different quantities which are shown to display an interesting behaviour in all the webs analysed. These quantities are computed on the *spanning trees* [13] of real webs and capture some important functional properties related to the degree of *efficiency* and *stability* of the networks. Moreover, we support our analysis by means of numerical simulations of the *Webworld Model* [19, 20], and show that these highly non-random properties can be reproduced by an evolutionary mechanism of Darwinian selection.

We finally discuss how these results shed new light on the theory of evolving networks, since the growth process differs radically from the traditional ones [5]. In particular, it is the result of an intrinsic coupling between network topology and the (population) dynamics defined on it.

8.2 Network Analysis

In this section we report the analysis of the data of four real food webs, namely those of St. Martin Island [8], Ythan Estuary with parasites [10], Silwood Park [9] and Little Rock Lake [6]. The latter three webs are the largest published food webs in the ecological literature, containing respectively 134, 154 and 182 species. We followed the common convention [1] of adding, if absent, a formal "environment vertex" representing the abiotic resources to the webs. Actually, the food web of Silwood Park documents all the interactions centred on the Scotch Broom *Cytisus scoparius*, which represents the "environment vertex" of the web. We report the analysis for both the unaggregated and aggregated versions of the webs.

8.2.1 Ecological Properties

Most of the traditional ecological quantities such as c_d, B, I, T, $(B+I)/(I+T)$ are computed in the original papers [6, 8, 9, 10] and are reported in Table 8.1. We also report the value l_{max} of the largest *trophic level* in the webs. The trophic level l of a species is defined as the minimum number of directed links separating it from the environment. It is a general result [1, 2] that the number of trophic levels of real food webs is always small even when the number of species is large. This is confirmed by looking at the values of l_{max} for the largest webs (see Table 8.1).

Table 8.1. Ecological properties of unaggregated (U.) and aggregated (A.) webs

	St.Martin Island		Ythan Estuary		Silwood Park		Little Rock Lake	
	U.	A.	U.	A.	U.	A.	U.	A.
S	44	42	134	123	154	82	182	93
L	224	211	597	576	365	215	2494	1046
c_d	0.11	0.12	0.03	0.04	0.01	0.03	0.07	0.12
B	0.14	0.14	0.04	0.04	0.12	0.23	0.34	0.13
I	0.70	0.72	0.57	0.57	0.11	0.17	0.65	0.86
T	0.16	0.14	0.39	0.39	0.77	0.60	0.01	0.01
$\frac{B+I}{I+T}$	0.97	1.00	0.64	0.64	0.27	0.52	1.52	1.14
l_{max}	4	4	4	4	3	3	3	3

8.2.2 Small-World Properties

As we already mentioned, the small-world effect (small value of the average distance D and large value of the clustering coefficient C_c) has been detected in a large number of different networks [5, 14]. Recent studies [11, 15, 16] have extended such analysis to food webs.

The behaviour of D in both unaggregated [15] and aggregated [11, 16] versions of the webs has been investigated. The remarkable result is that, even in the largest webs, the value of D is always less than or equal to 3 (this finding is obviously related to, and more general than, the aforementioned result concerning trophic levels). In Table 8.2 we report the value of D for the webs analysed.

Also the value of the clustering coefficient C_c of both unaggregated [15] and aggregated [11] food webs has been studied. While in some webs [15] the value of C_c is larger than that of a random graph with the same number of species S and connectance c, in other webs [11] the opposite is true. For the webs in our analysis (see Table 8.2), the value of C_c ranges from being 3.8 (Ythan Estuary) to 1.1 (St.Martin Island) times larger than displayed by random graphs [11].

In general, the variations observed in the behaviour of C_c mean that, differently from most real-world networks, food webs do not display small-world properties [11].

8.2.3 Degree Distribution

The degree distribution $P(k)$ displayed by food web has been studied by different authors [11, 15, 17]. In a study focusing on unaggragated webs [15] it was found that, while in some cases (such as Ythan Estuary and Silwood Park) the form of $P(k)$ can be fitted by a power-law, other webs (such as Little Rock Lake) display an irregular degree distribution. We performed a similar analysis on the aggregated webs and found the same behaviour as the unaggregated ones (see Fig. 8.1): the form of $P(k)$ is quite irregular for Little Rock Lake, while it can be roughly fitted by a power-law distribution ($P(k) \propto k^{-\gamma}$) for Ythan Estuary ($\gamma = 1.08 \pm 0.13$) and Silwood Park ($\gamma = 0.96 \pm 0.13$). Moreover, we found an irregular form of $P(k)$ for the St. Martin web too. Note that, however, the power-laws are very noisy.

Very recently, more comprehensive studies [11, 17] focusing on the *cumulative* degree distribution concluded that the form of $P(k)$ in most aggregated webs is

Table 8.2. Small-world properties of unaggregated (U.) and aggregated (A.) webs

	St.Martin Island		Ythan Estuary		Silwood Park		Little Rock Lake	
	U.	A.	U.	A.	U.	A.	U.	A.
D	1.93	1.92	2.41	2.40	3.34	3.06	2.15	1.89
C_c	0.32	0.31	0.22	0.23	0.14	0.23	0.38	0.54

152

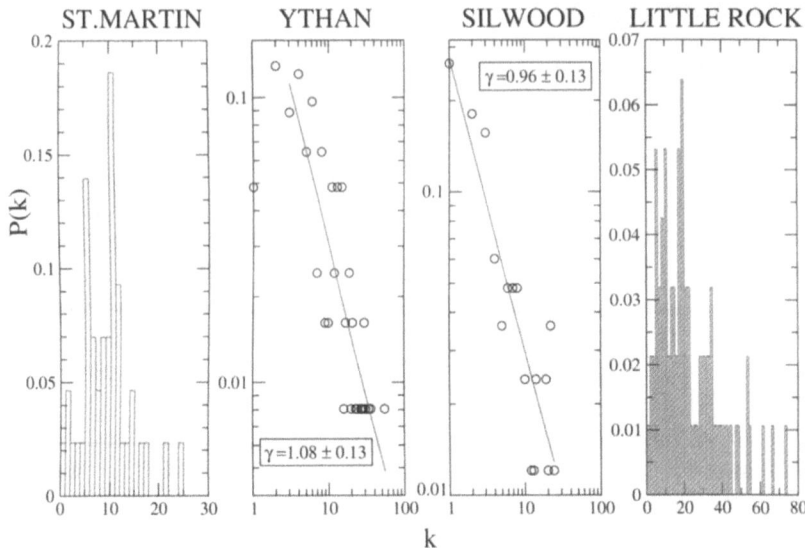

Fig. 8.1. Degree distribution $P(k)$ for the four webs analysed

not scale-free, but rather single-scaled. Thus, real food webs do not in general display neither small-world nor scale-free properties [11].

We anticipate here that the ambiguity in the form of $P(k)$ reflects a related difficulty in understanding food web organization. Scale-free networks can be reproduced by a simple *preferential attachment* [5] mechanism capturing the fundamental ingredient underlying their growth process: new vertices are continuously added and linked to preexisting ones with probability proportional to their degree. Instead, as regards food webs, preferential attachment is unlikely to be the correct growth hypothesis, for at least three reasons. First, it is not clear why the degree of the species should completely drive the evolution of the network [11, 17]. Second, the likelyhood of developing new reciprocal connections is determined by the features of the species, not simply by the number of their current interactions. Finally, food web evolution is a rather complex result of processes like speciation, extinction and rearrangement of interactions due to modifications in species' abilities [19, 20], an aspect which cannot be reduced to simple growth rules focusing only on topological quantities.

8.3 Spanning Tree Analysis

We have shown that, when looking at the traditional network properties of food webs, several difficulties arise. In the rest of the paper, we propose a different framework where food webs can be studied and modelled, and we suggest an analysis that allows to uncover unexpected regularities in real food webs. These ideas also help in modelling food web evolution in a more realistic way.

8.3.1 Foodwebs as Transportation Networks

The transfer of resources in a food web can be regarded as a transportation process starting from the environment and reaching, directly or indirectly, every species in the web. This is essentially due to the simple fact that every species has to be delivered a certain amount of resources (prey) to survive. In other words, the graph representing a food web is connected and such that each species can be reached starting from the environment and following the direction of the links.

Moreover, note that links in a food web are not simply binary (present or absent). Each link is indeed characterised by a "strenght" [1] which measures the amount of resources which is transferred in the predation it represents. An exact definition of link strenght which is also suitable for empirical observation is limited by conceptual difficulties [21]. However, it is generally accepted [1, 2] that the amount of resources transferred from a prey to a predator is small and such that each trophic level l delivers a fraction $\lambda \simeq 0.1$ of its resources to the level $l + 1$.

The connectedness of food webs implies that they have at least one directed *spanning tree*. A spanning tree of a directed graph with a "source" vertex (in our case, the environment) is a connected subgraph with no loops such that each vertex is reachable from the source. Moreover, the small value of the ecological efficiency λ implies that, roughly speaking, each species receives the largest amount of resources from the shortest sequence of links separating it from the environment. This means that, among all possible spanning trees of a food web, those representing the main transfer of resources are formed by the shortest chains from the environment to the species. For each food web in our analysis, we obtained a spanning tree with this features by firstly ordering the species in trophic levels and then removing all links directed from a prey at level l to a predator at level less than or equal to l. A few loops can still remain if more than one prey at level l supplies resources to the s ame predator at level $l + 1$. In this case, we randomly select only one incoming link for each species to obtain one spanning tree, and repeat this random procedure 1,000 times to have a set of equivalent spanning trees. Note that the "root" of the tree is the environment and the "leaves" are *top* species.

Once a spanning tree is obtained, one can analyse it in the context of transportation networks and extend to food webs what is commonly performed for other tree-like structures such as river basins [22] and vascular or respiratory systems [23, 24]. In particular, one can analyse how branching properties scale with system size (the number of vertices) according to the procedure described below.

8.3.2 Allometric Scaling Relations

River basins and vascular systems can be represented by tree-like graphs with S vertices labeled $i = 1, S$ plus an additional vertex $i = 0$ (the outlet of the basin, or the heart) which is the root of the tree. For each site i in the tree, compute

[18] the number A_i of vertices belonging to $\gamma(i)$ (defined as the set of vertices in the branch starting at i plus i itself) and the sum C_i of the quantities A_i in the same branch:

$$C_i = \sum_{j \in \gamma(i)} A_j \qquad (8.1)$$

(note that $\gamma(0)$ is the whole tree, and A_0 equals the total number of vertices plus the root $S + 1$). Plotting C_i versus A_i for each vertex i (including $i = 0$), allometric scaling relations of the form $C(A) \propto A^\eta$ are observed in both river networks [18] (with $\eta = 3/2$, A_i being proportional to the drainage area [22] uphill site i) and vascular systems [18, 24] (A_0 and C_0 are respectively proportional to the metabolic rate B of an organism and to its body mass M, which are empirically related [25] by $B \propto M^{3/4}$, yielding $\eta = 4/3$). The exponent η measures the efficiency [18, 24] of the transportation system in transferring resources from the root to the sites (as in vascular systems) or from the sites to the root (as in rivers), since C_i can be regarded as the "cost" of supporting the transfer, through $\gamma(i)$, of an amount of resources proportional to A_i. One can show [18, 24] that the most efficient topology for a tree-like network embedded in an Euclidean d-dimensional space corresponds to the optimal minimum exponent

$$\eta_{eff} = \frac{d+1}{d} \qquad (8.2)$$

(while the least efficient configuration, corresponding to a space-filling chain-like topology, yields the maximum value $\eta = 2$). Hence, both vascular ($d = 3$) and river ($d = 2$) networks are optimized, a result revealing their highly non-random organization and thus being a signature of the (biological [23, 24] or hydrogeological [22]) evolution that shaped their structure.

8.3.3 Efficiency of Empirical Food Webs

We computed the quantities $C(A)$ on the spanning trees of both unaggregated and aggregated food webs. In Fig. 8.2 we show the result of such analysis once an average over the set of 1,000 spanning trees of each web is performed. Remarkably, all webs display a clear power-law behaviour of the form $C(A) \propto \eta$.

The exponent η is always in the range $1.11 - 1.25$, which reveals a high degree of efficiency in the webs (the exponent is smaller than in rivers and organisms). This is because food webs are not embedded in any Euclidean space, so that there is no dimension d constraining the webs to display an exponent larger than the optimal value η_{eff} previously discussed. The most efficient configuration that can be realised without such constraints is the *star-like* topology where all species are directly connected to the environment. Note that $\eta \to 1$ as a generic tree approaches a star-like configuration. At the opposite limit, as we already mentioned, one has a *chain-like* configuration with $\eta = 2$. The finding that η is much closer to 1 than to 2 is thus related to the small number of trophic levels

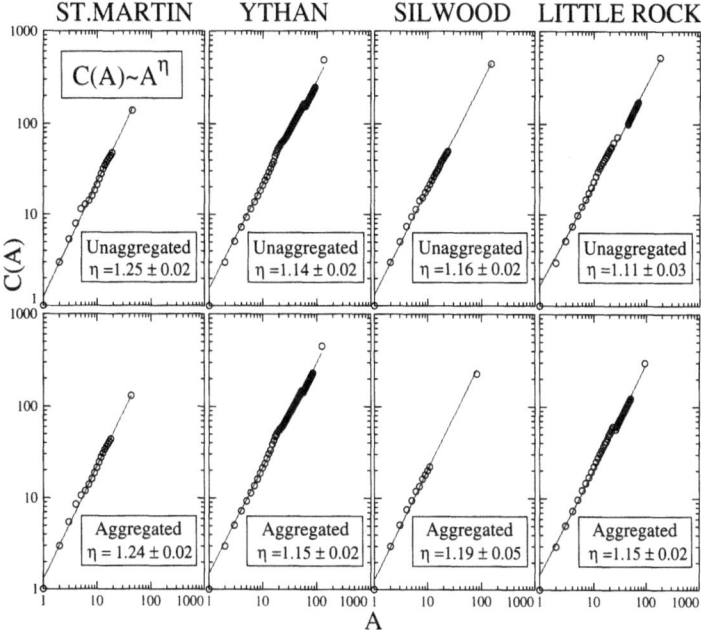

Fig. 8.2. Allometric scaling relations $C(A)$ for the four webs analysed

in the webs (in the star-like configuration there is only one trophic level, while in the chain-like one there are S distinct levels).

We note that aggregated webs display a better form of $C(A)$ than the unaggregated ones, and also that in the three largest webs the values of η are all consistent with each other. We suggest that the larger value displayed by St. Martin is due to the small size of the network (the points with small values of A tend to increase the value of η with respect to the large-scale behaviour; such points have a stronger effect in small webs). It would be interesting to include more webs in the analysis to test whether the large-scale behaviour of the networks is universal [26].

The most striking result is that the power-law form of $C(A)$ means that, like rivers [22], the topology of any branch of the tree is statistically equivalent to that of the whole tree. The finding of such *self-similarity* is remarkable in the ecological context, since it might suggest that the overall organization of the web is the result of local processes shaping the same form of the network at all scales. It is thus important to ask what processes are suitable candidates for the emergence of such pattern. We suggest that a possible answer is that every species chooses a trade-off between maximizing resource input (by preying on species at the lowest possible level) and minimizing the effort to compete against the other predators which have the same aim. This results in a highly efficient, although not optimal ($\eta = 1$) topology of the network. For these reasons, we expect that a suitable model including competition effects might reproduce the observed form of the scaling relation $C(A)$.

Moreover we remark that, since communities evolve (new species add and old ones disappear), the number of species to support is not fixed *a priori*, but rather the result of the best arrangement of a growing number of species among trophic levels. Following the analogy with metabolism, it would be then very interesting to relate the quantity C_0 (which throughout this paper will be simply treated as a topological quantity) to the total amount of resources needed to support the whole system. With this observations, probably the correct way of regarding the optimization of the food web is reversed: one should not ask whether C_0 (viewed as a "cost") is minimized once $A_0 = S + 1$ is fixed, but rather whether A_0 (the diversity of the community) is maximized once C_0 (viewed as a measure of the environmental supply) is fixed. These issues are of fundamental interest in biogeography [27].

8.3.4 Stability under Species Removal

Until now, we did not discuss the role of the links that are eliminated to obtain the spanning trees. While (according to our discussion) they have no relevance in determining network efficiency, we note that such links are likely to be essential to the *stability* of the webs. To see this, we considered the statistical distribution $P(A)$ of the branch size A in one of the possible spanning trees of each food web (see Fig. 8.3). In each case, the form of $P(A)$ is higly skewed.

Now, note that if the web coincided with its spanning tree then $P(A)$ would measure the probability that the elimination of a randomly selected species results in the consequent removal of A species. The skewed behaviour of $P(A)$ means that there is a large number of species whose removal would result in the elimination of few species, while the single removal of one of a small number of other species would dramatically affect the stability of the web. In the whole network, instead, the presence of additional links ensures a larger stability. To test this property, we artificially removed one randomly selected species from the original webs and repeated this test for each species in all webs, finding that *the maximum fraction f of species being eliminated after one random removal is $f \simeq 0.05$, with no regular dependence on the degree of the removed species*. This proves a very high stability of the food webs, which is not due to the presence of the species with the largest degree. This is an a dditional argument related to the form of the degree distribution of food webs: while in scale-free networks the vertices with the largest degree are responsible for the connectedness of the whole network [5], in food webs this is not the case. Rather, food web stability relies on the large number of "redundant" links (those absent in the spanning tree), as confirmed by observing (see Table 8.1) that in all webs $L \gg S$ (while in a tree $L = S$).

In some webs the form of $P(A)$ can be roughly approximated by a power law of the form $P(A) \propto A^{-\tau}$ (see Fig. 8.3). A power-law behaviour of such quantity is widely observed in river networks [22], where $P(A)$ is the *drainage area distri-*

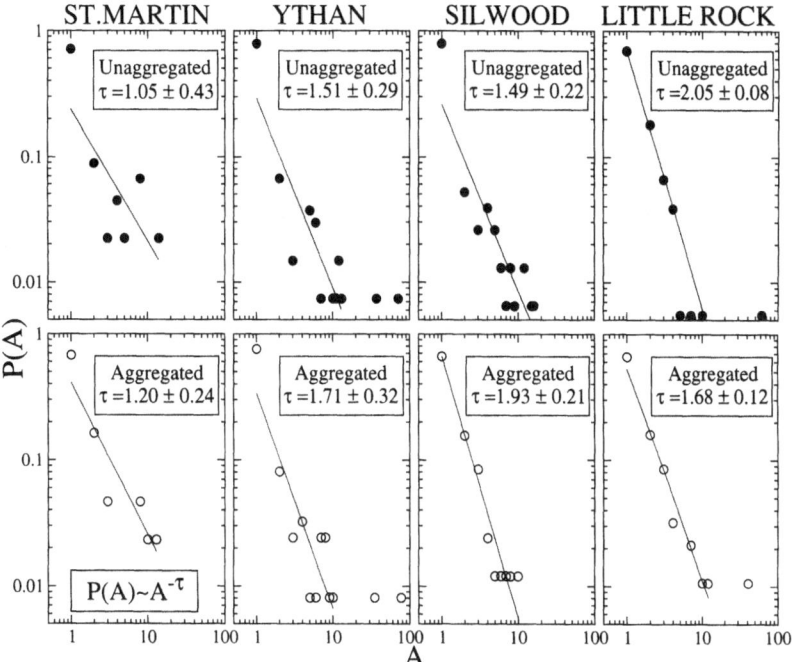

Fig. 8.3. Branch size distribution $P(A)$ for the four webs analysed

bution. Here, the agreement with the fitting curve is better for aggregated webs, however the variations in the value of the exponent are such that we shall not consider τ as a significant quantity in our following analyses of food web models. For the same reason, we shall not consider the behaviour of $P(k)$, and focus only on the values of the connectance c (which allows comparison with the models) and the scaling exponent η (which is the relevant newly introduced quantity). Finally, we shall only consider aggregated webs.

8.4 The Webworld Model

Perhaps the most relevant models of food web structure are the *Cascade model* [3] and the more recent *Niche model* [7]. Both models are *static*, since they generate webs with a fixed number of species and the desired value of the connectance (externally tuned by a control parameter). The Niche model improves the predictions of the Cascade model and reproduces several food web properties [7]. However, static models are of limited relevance in understanding the structure of real food webs, since they do not explicitly highlight any organising principle. More specifically, due to the above considerations on the efficiency and stability properties of real food webs, the processes shaping the observed patterns are likely to be strongly related to the *evolution* [19] of the networks themselves.

8.4.1 Coupling between Topology and Dynamics

To capture such aspect, we present the numerical results of the Webworld model [19, 20], which simulates food web evolution under the long time processes of speciation and extinction. Such evolving character of the model makes it significantly more complex than static ones, however the basic idea is simple. Since the modifications in the topology of the network are due to the introduction of new species and the elimination of old ones, and since such processes cleary depend on the *population dynamics* defined among species, the model explicitly takes the dynamical aspect into account. This results in the definition of species in terms of some features that determine the interaction coefficients of a set of population equations governing the number of individuals of each species. Variations in network composition allows species to explore the set of possible features, so that the food web progressively evolves from a random to a complex topology.

Before giving a more detailed description of the model and its predictions, we note here that in such growth process the knowledge of the topology of the web at a certain timestep is not sufficient in order to simulate (even stochastically) the topology at the following timestep. The additional knowledge of dynamical variables (the population and features of the species) is essential. This aspect is absent in all models based on the *preferential attachment* hypothesis or variations of it [5], where the topology of a network at a timestep is obtained by means of stochastic rules once the topology at the previous timestep is known.

8.4.2 Initial State of the Model

All versions [19, 20] of the Webworld model define species in terms of a set of J phenotypical features (picked from a pool of K possible ones) that can change in time, and differ in the form of the population dynamics. The environment is treated as an additional species $i = 0$ and assigned a set of J features that do not change in time. The initial number of species S is an arbitrary choice, since the long-term properties of the model do not depend on it [19, 20]. We will refer to the early version of the Webworld model [19] since, as we will show, it displays the minimal ingredients yielding the observed features.

Potential Predation Scores

The usefulness of possessing the feature α when predating a species possessing the feature β is given by the element m of a $K \times K$ anti-symmetrical random matrix whose elements are picked uniformly in the interval $[-1, 1]$ (in our simulations, according to the original work [19], we always set $K = 500$ and $J = 10$). A species i has a total predation "score" against species j given by $S_{ij} = max\{0, \sum_{\alpha\beta} m_{\alpha\beta}/J^2\}$, where α runs over the set of features of i and β runs over the set of features of j, so that $0 \leq S_{ij} \leq 1$ (actually, in the original paper [19] the elements m_{ij} are picked from a Gaussian distribution with zero mean and unit variance, and the sum $\sum_{\alpha\beta} m_{\alpha\beta}$ is divided by J, not by J^2, to have a unit variance of S_{ij}; our modification is unessential and simplifies the analytical predictions). A positive value of S_{ij} means that i is potentially adapted to be a predator of j, while j cannot be a predator of i (the environment can only be predated, thus $S_{0i} = 0$ for each i).

Competition

To be an effective predator of j, species i has to compete with the other predators of j. More specifically, its score S_{ij} has to be greater than the threshold value $S_j^M - \delta$, where $S_j^M = max_i\{S_{ij}\}$ is the score of the *main predator* of j and $0 \le \delta \le 1$ is a parameter of the model determining the strength of competition. Equivalently, in order to draw a link from j to i in the food web, the "effective" score

$$F_{ij} = max\{0, 1 - (S_j^M - S_{ij})/\delta\} \qquad (8.3)$$

has to be positive. A larger value of δ (weaker competition) means that more species are effective predators, hence δ determines the connectance of the network (note that cannibalistic and double loops are not allowed [19]): when $\delta = 1$ (minimum competition) all potential predators are effective (a link exists between any pair of species), while when $\delta = 0$ only main predators are allowed (the food web reduces to a chain), thus δ is analogous to the tuning parameter of static models.

Initial Properties of the Model Food Webs

Before discussing the evolution of the model, we first analyse the initial state of the webs. In the following, we will always compare the *aggregated* versions of both real and model webs. However, since in the model features are randomly assigned, the probability of finding two trophically equivalent species is vanishing in the initial step.

As we mentioned, the initial connectance c^0 strongly depends on the value of the competition parameter δ. The general expression relating δ and c^0 in the initial state of the model can be obtained by noting that the total number of links L must equal the sum of the number of predators (k_{out}) of each species. For each species j there are $S-1$ species i having scores S_{ij} against it, of which on average $(S-1)/2$ are zero, while the remaining $(S-1)/2$ are uniformly distributed in the interval $[0, 1]$. The total number of predators of j is thus given by 1 (the main predator) plus the fraction of the remaining $(S-1)/2 - 1 = (S-3)/2$ scores falling within the segment of length δ, which is $k_{out} = 1 + \delta(S-3)/2$. Hence the connectance c^0 is given by

$$c^0 = 2L/S^2 = (2/S^2) \sum k_{out} = 2/S + \delta(S-3)/S \simeq 2/S + \delta \qquad (8.4)$$

Table 8.3. Properties of the initial state of the Webworld model with 1,000 species

δ	c_d^0	η^0
0	0.001	2.00
0.01	0.005	1.83
0.02	0.01	1.51
0.1	0.05	1.33
0.2	0.1	1.22
0.6	0.3	1.18
1.0	0.5	1.05

Note that this expression reduces to the correct limiting values when $\delta = 0$ ($c^0 \simeq 2/S$) and $\delta = 1$ ($c^0 \simeq 1$). Equivalently, the *directed connectance* $c_d^0 \simeq c^0/2$ has the initial value $c_d^0 \simeq 1/S + \delta/2$. By looking at Table 8.3, we see that this analytical prediction is confirmed by the numerical results. We note that the values of the directed connectance of real webs (see also Table 8.1) are reproduced by setting δ to a value between 0.02 and 0.2.

The value η^0 of the scaling exponent for the webs generated in the initial step of the model is also reported in Table 8.3. The values range from 2 to 1 as δ varies from 0 to 1. This means that, as expected, the spanning trees of the webs range from the chain-like to the star-like configuration as δ increases. However, to have a value of η^0 within the observed range (see Fig. 8.2) the competition parameter has to be set to a value $\delta \geq 0.2$, corresponding to values of the connectance which are too large (see Table 8.3). Therefore, the values of the connectance and of the scaling exponent cannot be simultaneously reproduced with a single choice of δ, and the initial state of the Webworld model is unsuitable for generating realistic webs.

8.4.3 Evolution of the Model

The incompatibility of the values of the connectance and of the scaling exponent in the initial state of the model can be regarded as follows. To have a spanning tree close to a star-like topology (like those of real food webs), the number of links in a randomly assigned web has to be large, so that there is a large probability for any species to be connected to the environment vertex or to first level species. In real webs, instead, even with a smaller number of links the topology is close to optimality. This is likely to be the result of the *evolution* of real webs, which were shaped by *local* processes increasing the *global* efficiency of the system, in the same way that the optimized topology of river basins and vascular systems is the result of a nontrivial evolutionary mechanism [18, 22, 23, 24].

To test these predictions, we now introduce the rules governing the evolution of the Webworld model [19] and discuss their results.

Population Dynamics

The scores F_{ij} determine the coefficients γ_{ij} of the population dynamics governing the number of individuals of each species i, which are assumed to display the stationary values N_i given by the following simple set of linear *donor-controlled* [28] equations:

$$N_i = \lambda \sum_j \gamma_{ij} N_j \qquad (8.5)$$

where λ is the *ecological efficiency*, which is set to the value 0.1 consistently with the empirical estimates [2], j runs over all species (including the environment, which supplies a constant amount of resources $R = \lambda N_0$) and the coefficients

$$\gamma_{ij} = \frac{F_{ij}}{\sum_k F_{kj}} \qquad (8.6)$$

are the normalized "effective" scores (the coefficient γ_{ii} is defined to be 0 or -1 if i has respectively no predators or at least one).

Speciation and Extinction

At each timestep t, all species with $N_i < 1$ (less than one individual) are removed from the web (*extinction event*), while a new species is added (*speciation event*), differing in only one randomly chosen feature from a pre-existing species (chosen with probability proportional to its population N_i). Then the new stationary population sizes are computed, and so on.

The newly introduced species may be unfit to compete against the preexisting ones, and it can go extinct at the end of the following timestep. Otherwise, it can successfully add to the web in a stable way, and even cause the extinction of its competitors. After enough time, such evolutionary processes select successful features and reject disadvantaegous ones, so that the list of features of each species is no longer random. Consequently, the scores determining the topology of the web form highly correlated patterns and the complexity of the network increases significantly.

Asymptotic State of the Model

The result of a typical simulation of the model is reported in Fig. 8.4 (with $R = 900$ and $\delta = 0.04$). We stress again that the model webs have to be aggregated before computing the quantities of interest. The evolution of the network is monitored recording, every $5,000$ timesteps, the number of species $S(t)$, the scaling exponent $\eta(t)$ (computed by an internal procedure of the program fitting the relation $C(A)$ with a power law) and the connectance c.

All quantities display an approximately asymptotic behaviour after an initial stage of evolution. During this initial stage, the number of species $S(t)$ grows,

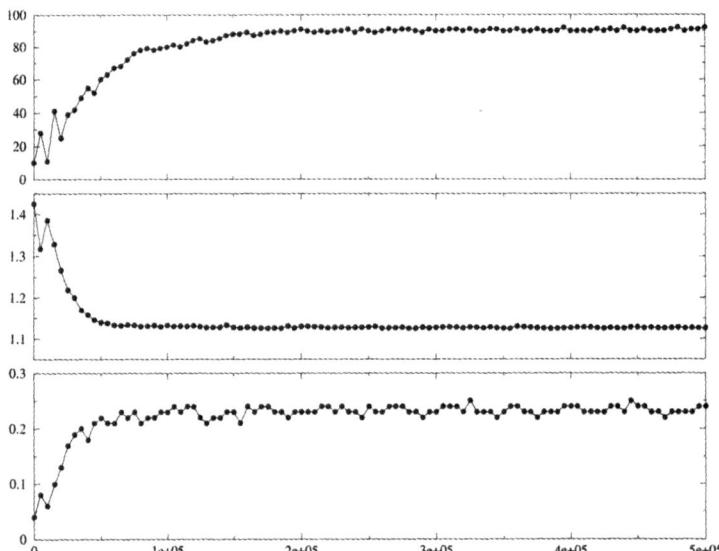

Fig. 8.4. Evolution of a simulation of the Webworld model with $R = 900$ and $\delta = 0.04$ (the abscissa is the time axis in units of the number of timesteps t). Upper panel: evolution of the number of species $S(t)$. Middle panel: evolution of the scaling exponent $\eta(t)$. Lower panel: evolution of the connectance parameter $c(t)$

punctuated by sudden decays ("coevolutionary avalanches") since species are not yet adapted to coexist in a stable way [19]. The fluctuations in the number of species are thus large at the beginning and decrease as time proceeds. A similar trend, namely the decrease in the rate of extinction (number of species families eliminated per unit time) is documented in the Fossil Record [29], a finding which has been interpreted [30] as a progressive increase in the degree of adaptation of the species to coexist in the environment. The number of species then saturates to an almost constant value, however species continue to speciate, so that the composition of the web changes despite the diversity is approximately constant. Once δ is fixed, the final number of species increases as R increases, independently of the initial value of S.

Evolution of the Scaling Exponent

More interestingly (see Fig. 8.4), the exponent η decreases in time towards an asymptotic value η^∞ (1.13 in the figure) which is reached much before ($t \simeq 100,000$) the number of species sets to its stable value ($t \simeq 200,000$), so during a long time interval the web grows while η remains constant showing its independence on system size. The decrease of η during the evolution has a clear interpretation in terms of the efficiency of the food web. Starting from a randomly assigned (hence inefficient) initial state, the topology of the web evolves through successive "local" events (speciation and extinction) driven by the

Table 8.4. Properties of the asymptotic state of the Webworld model with $R = 1,000$

δ	c_d^∞	η^∞
0	0.01	2.00
0.01	0.08	1.61
0.02	0.10	1.20
0.1	0.12	1.11
0.2	0.18	1.09
0.6	0.37	1.07
1.0	0.50	1.04

coupling with the dynamics defined on it, thus becoming more and more globally efficient over time.

Once R is fixed, the asymptotic value η^∞ depends on δ (see Table 8.4) and, as expected, it ranges from $\eta^\infty = 2$ ($\delta = 0$) to $\eta^\infty = 1$ ($\eta = 1$). We checked that the dependence on R is such that η^∞ increases slightly even when R is increased over orders of magnitude (yielding a final number of species $S \gg 1,000$, falling out of the testable range). Hence, as a rule of thumb, R determines the final value of S (and has to range between 200 and 1,000 to have from 50 to 150 final species) and δ determines the scaling exponent η^∞. The asymptotic values of η correspond to the most efficient topology that the network can reach for a given value of δ, this optimized state being significantly more efficient than the initial (random) configuration.

Evolution of the Connectance Parameter

The connectance c of the web (see Fig. 8.4) increases during the first stage of the evolution, reaching (for the choice of the parameters in the figure) the asymptotic value $c^\infty = 0.23$ (corresponding to $c_d^\infty = 0.12$) which is almost six times larger than its initial value $c = \delta = 0.04$. In Table 8.4 we report the asymptotic values c_d^∞ corresponding to various choices of δ.

The increase in the connectance can be explained by noting that "least fit" (low-score) species are removed while high-score species are replicated and mutated, hence after enough time most scores will be found above the threshold value imposed by δ increasing the network connectance. This means that the species surviving to the continuous modification in the network topology (due to the evolution of the web) are those selected to have many links.

Note that the evolution of the network is such that, differently from the initial state, the asymptotic configuration fits both the connectance and the scaling exponent of at least the webs of St. Martin and Little Rock Lake. This occurs when δ is set to an appropriate value in the range $0.02 \leq \delta \leq 0.1$. In particular, the choice of the parameters corresponding to the simulation shown

Table 8.5. Comparison between the web of Little Rock Lake and the asymptotic state of the Webworld model with $R = 900$ and $\delta = 0.04$

	Little Rock Lake	Webworld Model
S	93	93
L	1046	1037
c_d	0.12	0.12
B	0.13	0.15
I	0.86	0.84
T	0.01	0.01
$\frac{B+I}{I+T}$	1.14	1.16
l_{max}	3	3
D	1.89	1.89
C_c	0.54	0.55
η	1.15	1.13
τ	1.68	1.72

in Fig. 8.4 was aimed at reproducing the web of Little Rock Lake. A more detailed comparison between the simulated and the real web is reported in Table 8.5 (the match is excellent). This suggests that the model captures the key evolutionary principles underlying the organization of real communities.

8.4.4 Further Comments on the Model

Despite the Webworld model looks quite complicated, once λ is fixed to its phenomenological value the only relevant parameters are R (determining the final number of species) and δ (tuning all other topological quantities such as η and c), in the same way that simpler static models require the number of species and another parameter (the connectance or an equivalent one) to determine the same topological properties. The freedom in choosing the value of R corresponds to the fact that different real food webs are supported by different environments, each supplying a certain amount of resources. For the same reason, the number of observed species varies significantly across real food webs.

As regards the parameter δ, instead, we comment that while in the model it has to be externally tuned to a proper value to yield the desired value of the exponent η, it is important to ask what natural mechanism may be responsible for the selection of a particular value of η in real food webs. We propose two explanations.

The first is a simple hypothesis of the existence of a negative feedback between the amount of exploitable resources (in the model, R) and the strength of competition (δ): when resources are abundant (or not completely exploited), the competition among species is negligible, and the number of species in the first trophic level can grow. But when the amount of resources is decreased, the competition among first-level species increases, so that new species find a "better" ecological niche by feeding on species at first level, hence starting to occupy higher levels. This feedback process might set both quantities to equilibrium values corresponding to the observed exponent η.

The second is the hypothesis of a self-organized [31, 32] scenario in which the predation threshold (δ in the model) spontaneously sets to a critical value with no external fine tuning. While in the Webworld model the updating of predation scores (through speciation and extinction) depends on the input value of δ, the real processes changing the predation abilities of species may be self-organized and result in an updating algorithm similar to that defined in the Bak–Sneppen model [31] (species are arranged in a food chain and assigned a fitness value uniform in the interval $[0, 1]$; if at each timestep a new fitness value is assigned to the species with the minimum value and to its nearest neighbours, after enough time all fitness values will be found above a threshold which spontaneously sets to a particular value), yielding the observed value of η with no fine tuning of any control parameter.

8.5 Discussion and Conclusion

We showed that real food webs display previously undetected properties that are likely to be related to their evolution. The underlying growth process is such that real webs show a high degree of both efficiency and stability. Such properties suggest that food web modelling cannot ignore the evolutionary aspects leading to a highly non-random network organization. This is confirmed by comparing the features of the initial state of the Webworld model to its "asymptotic" behaviour. Indeed, Darwinian evolution increases the efficiency and the complexity of the model food webs. While the (random) initial state of the model is not able to reproduce real food webs, the (highly structured) asymptotic state succeeds in doing so.

Food webs are not the only example of real-world networks displaying a complex topology which is not reproduced by simple static models. As we mentioned, one of the possible mechanisms leading to a scale-free degree distribution is the *preferential attachment* hypothesis. However, we suggested that while a similar hypothesis is justified in the case of social networks (where the evolution can indeed be driven by network topology alone), in food webs this is not the case. The additional introduction of a set of *dynamical* variables (the population sizes of the species) seems necessary.

Since their coherent definition in terms of directed graphs, the study of food webs has benefited from the advances in the field of complex networks, new aspects of their structure having emerged by comparison with other different systems. We think that the concepts discussed here may establish a feedback on this knowledge process, since food webs represent a prototype example of evolving networks whose topology and dynamics are tightly coupled, an aspect that has been so far ignored in the modelling of complex networks and, as we showed, requires deeper insight into their structure. Since this additional degree of complexity is widespread in real-world networks, the understanding of food web evolution may be the starting point for a more general interdisciplinary exploration of this intriguing subject.

References

1. J.H. Lawton: 'Food Webs'. In: *Ecological Concepts.* ed. by J.M. Cherret (Blackwell Scientific, Oxford 1989) pp. 43–48
2. S.L. Pimm: *Food Webs* (Chapman & Hall, London 1982)
3. J.E. Cohen, F. Briand, C.M. Newman: *Community Food Webs: Data and Theory* (Springer, Berlin 1990)
4. S.H. Strogatz: Nature **410**, 268 (2001)
5. R. Albert, A.-L. Barabási: Rev. Mod. Phys. **74**, 47 (2002)
6. N.D. Martinez: Ecol. Monogr. **61**, 367 (1991)
7. R.J. Williams, N.D. Martinez: Nature **404**, 180 (2000)
8. L. Goldwasser, J. Roughgarden: Ecology **74**, 1216 (1993)
9. J. Memmott, N.D. Martinez, J.E. Cohen: J. Anim. Ecol. **69**, 1 (2000)
10. M. Huxham, S. Beaney, D. Raffaelli: Oikos **76**, 284 (1996)
11. J.A. Dunne, R.J. Williams, N.D. Martinez: Santa Fe Institute Working Paper 02-03-10 (2002)
12. N.D. Martinez: Am. Nat. **139**, 1208 (1992)
13. B. Bollobás: *Random Graphs* (Academic, London 1985)
14. D.J. Watts, S.H. Strogatz: Nature **393**, 440 (1998)
15. J.M. Montoya, R.V. Solé: J. Theor. Biol. **214**, 405 (2002)
16. R.J. Williams, N.D. Martinez, E.L. Berlow, J.A. Dunne, A.-L. Barabási: Santa Fe Institute Working Paper 01-07-036 (2001)
17. J. Camacho, R. Guimerà, L.A.N. Amaral: Phys. Rev. Lett. **88**, 228102 (2002)
18. J.R. Banavar, A. Maritan, A. Rinaldo: Nature **399**, 130 (1999)
19. G. Caldarelli, P. G. Higgs, A.J. McKane: J. Theor. Biol. **193**, 345 (1998)
20. B. Drossel, P.G. Higgs, A.J. McKane: J. Theor. Biol. **208**, 91 (2001)
21. M.S. Laska, J.T. Wootton: Ecology **79**, 461 (1998)
22. I. Rodriguez-Iturbe, A. Rinaldo: *Fractal River Basins: Chance and Self-Organization* (Cambridge University Press, Cambridge 1996)
23. G.B. West, J.H. Brown, B.J.A. Enquist: Science **276**, 122 (1997)
24. G.B. West, J.H. Brown, B.J.A. Enquist: Science **284**, 1677 (1999)
25. T.A. McMahon, J.T. Bonner: *On Size and Life* (Scientific American Library, New York 1983)
26. D. Garlaschelli, G. Caldarelli, L. Pietronero: in preparation (2002)
27. R.H. MacArthur, E.O. Wilson: *The Theory of Island Biogeography* (Princeton University Press, Princeton 1967)
28. D.W. Zheng, J. Bengtsson, G.L. Agren: Am. Nat. **149**, 125 (1997)
29. J.J. Sepkoski Jr: Paleobiology **19**, 43 (1993)
30. P. Sibani, M.R. Schmidt, P. Alstrø m: Phys. Rev. Lett. **75**, 2055 (1995)
31. P. Bak, K. Sneppen: Phys. Rev. Lett. **71**, 4083 (1993)
32. H.J. Jensen: *Self-Organized Criticality* (Cambridge University Press, Cambridge 1998)

9 Social Networks: From Sexual Networks to Threatened Networks

H. Eugene Stanley[1] and Shlomo Havlin[2]

[1] Center for Polymer Studies and Department of Physics
   Boston University, Boston, MA 02215, USA
[2] Minerva Center and Department of Physics
   Bar-Ilan University, Ramat Gan, Israel

Abstract. Our scientific goal is to uncover common principles governing the behavior of a range of social networks. Our practical goal is to use this understanding to develop specific strategies to destroy threat networks and, in parallel, to develop specific strategies to defend threatened social networks against attack. There are recent hints that progress toward achieving both goals can be achieved applying new approaches from modern statistical physics to social network structure and dynamics.

9.1 Introduction

Populations, which can be viewed as networks of social acquaintances, are vulnerable to disease epidemics such as AIDS. Any random immunization of people against such disease attacks is problematic because it must encompass almost the entire population in order to successfully stop the spreading epidemic [1, 2, 3, 4, 5]. Other types of social networks are organizations, e.g., security agencies, in which working relations are represented by links. To be effective, these organizations must be stable and allow rapid data flow in the network. We have begun addressing these problems – using concepts and tools of both social sciences and statistical and nonlinear physics – by designing more stable social network structures, enabling them to resist both random and intentional attacks. For this purpose, we need to better understand the topological structures of existing social networks, and to improve our understanding of transport in such systems.

Our methods in statistical physics are based on relatively new concepts, such as correlated site-bond percolation theory [6, 7, 8, 9, 10]. The applications of percolation theory range from predicting the amount of oil that can be extracted from an underground reservoir, to understanding the network formation mechanism involved in the hardening of a boiled egg. The use of percolation theory has already proven valuable in the study of social networks. The Bar-Ilan group has generalized percolation theory in order to analyze the structure and stability of general networks under random failures [11] and intentional attacks [12]. Based on this generalization, we are following up on a novel approach for designing new social networks that are more resilient to attack. We are also developing methods based on the percolation approach [13] that will enable us to immunize populations more effectively against different types of epidemics.

9.2 Recent Advances on Scale-Free Social Networks

Very recent analysis of social networks, as well as many other networks (such as trust networks and sexual networks), reveals that some of these networks display the important property of being scale-free [6, 2, 14], i.e., there is a very wide distribution of the number of links per vertex. Most vertices have a small number of connections. However, there are a small number of vertices that have a very large number of connections, and there are vertices in the full range between these extremes. Further, it seems that there is a possible explanation for this scale-free behavior [2, 15], and that the results for sexual networks extend to other social networks [16].

Our groups are studying the structure of a wide range of social network types [17], and are building mathematical models and tools for large social networks [13]. In studies conducted about the stability of scale-free social networks, it was proven that these networks are optimally resilient to the random failure of individuals [11]. Even if almost all elements of a network malfunction, a large fraction of the individuals will be left connected, allowing continuing interactions between a large fraction of the population. This situation is unlike that of homogeneous networks, in which such a failure will break the entire network into small, unconnected islands. On the other hand, a deliberate, successful attack on the most-connected elements in the network will lead to failure of the entire network after only a small fraction of nodes have been targeted [12]. Further, studies show that search can be conducted in such heterogeneous networks in a much more efficient way than in homogeneous networks [18].

A connection exists between (a) the stability of a network and (b) the propagation of disease. Heterogeneous networks are prone to the rapid spread of epidemics. If the individuals to be immunized are chosen randomly, spreading is unavoidable, even if almost all individuals in the network are immunized. However, if the individuals to be immunized are chosen using "smart" strategies, it becomes possible to reduce the number of infected individuals to almost zero. Using models, it is possible to forecast the consequences of epidemic outbursts and to try to control them. It is established that random immunization of a large fraction of the population fails to prevent epidemics of diseases that spread upon contact between infected individuals; for example, Malaria requires 99% of the population to be immunized in order to stop epidemic spreading [4, 5]. On the other hand, targeted immunization of the most-connected individuals requires global knowledge of the topology of the social network in question, rendering 99% immunization impractical. We recently proposed an effective strategy, based on the immunization of a small fraction of *acquaintances* of randomly-selected individuals, that prevents epidemics without requiring global knowledge of the social network [19].

9.3 Recent Advances on Traffic Flow in Networks

We are adapting recent results on traffic flow to social network analysis. In 1994, Leland et al. [20] found that Ethernet LAN traffic is self-similar; "bursts" occur on every time scale. These findings show that long-range correlations in the interval times of arriving packets and extreme variability (or infinite limit of the variance). Paxson and Floyd [21] have found evidence for self-similarity of Wide Area Network (WAN) Traffic, and showed the failure of Poisson modeling in this case. New empirical findings challenge the validity of the traditional queuing models, and new models have since been proposed. In contrast to the above measurements, Takayasu et al. [22, 23, 24] have measured a $1/f$ power spectrum only at the critical point of a phase transition, and it is still not clear whether the flow is always self-similar in such networks. They found finite correlation times in the fluctuations of network traffic, and identified phase transitions between "sparse" and "jam" phases of the network.

The empirical phenomena mentioned above can influence the design of control schemes for traffic. However, the empirical description of the traffic is not yet complete. As the Bar-Ilan group has demonstrated recently in the case of vehicular traffic [25], a careful nonlinear statistical analysis of measured data may lead to the finding of several congested phases. One of our goals is to clarify this issue, and one method that we will use in the analysis of measured time series is Detrended Fluctuation Analysis (DFA). DFA was developed by the Boston group [26] and has been successfully applied by us and others to many systems, e.g., to DNA sequences [27, 28], the analysis of climate changes [29, 30], heart rate variability [31, 32, 33, 34], economics [35], and even prime numbers [36]. One of the advantages of this method is its ability to detect long-range correlations in the records in the presence of trends and other nonstationarities.

9.4 Characteristic Properies of Real Networks

9.4.1 Classification of Real Networks

We have developed a method that classifies complex real-world networks according to their statistical topological properties [17]. By studying a wide range of different types of networks, we find evidence for the occurrence of three classes of small-world networks:

(a) scale-free networks,
(b) broad-scale networks, characterized by a connectivity distribution that has a power-law regime followed by a sharp cut-off;
(c) single-scale networks, characterized by a connectivity distribution with a fast-decaying tail.

9.4.2 Percolation

A percolation approach for general networks has been developed, with surprising results for scale-free networks [11, 12, 13]. The network is fully resilient to the random failure of sites and is extremely vulnerable to intentional attack. This analytical approach is being developed to study realistic social networks – e.g., where known correlations between individuals are included – where the measured clustering property and real geographical distance, measured experimentally, are being taken into account. Preliminary findings show that the geographical effect has a strong influence on the stability and transport of the network [37, 38, 39].

9.4.3 Structural and Transport Properties of Networks

We are studying several topological properties of networks – e.g., clustering and correlations. Some preliminary results already exist, such as the work on clustering in trust networks [40]. The clustering coefficient [41, 42], which quantifies the extent to which nodes adjacent to a given node are linked, seems not to be affected when the network collapses. This may be relevant to terrorist organizations that are comprised of small, strongly-connected cells that are connected to each other by a few, highly-connected individuals [43]. The clustering was found to be important also in electric power networks, e.g., the power grid in the Western States in which the clustering coefficient is significantly larger than that of random networks. A useful method to quantify correlations (by measuring assortative tendencies, i.e., the tendency of high-degree vertices to associate preferentially with other high-degree vertices) was suggested recently by Newman [44].

We have preliminary results extending these studies to other real social networks. We are also studying the degree distribution for sites at a given distance from the most-connected site [45]. We are also studying the effect of geographical distance in real networks. This information is important for evaluating the stability and the immunization threshold. We are also analyzing the transport properties of data flow in social networks. We are applying DFA analysis and multifractal analysis [46] to better understand transport in complex social networks. We also are developing structural and transport modeling that will enable a better understanding of the structure and transport in such networks.

9.4.4 Optimizing the Stability of Threatened Networks

We are using the analytical approach we developed to calculate the percolation threshold for a given network [11, 12], in order to design topologies that improve the stability of scale-free networks under both random failures and intentional attacks. This is being done by calculating the percolation threshold while keeping the average number of links for an individual in the network constant (for safety and security reasons) and then varying parameters such as the form of the degree distribution, the type of correlations, and the clustering coefficients. We are also testing the effect of geographical distances on the stability of scale free networks. This will enable us to propose ways to design more stable networks and to improve the stability of existing networks.

9.4.5 Immunization of Networks

Random immunization fails to prevent epidemics of diseases that spread in populations upon contact between infected individuals [4, 5]; the same is true for immunization of computers against viruses [47]. Unless almost the entire system is immunized, the virus continues to spread through the population or computer network. To deal with this problem, the Bar-Ilan group has developed an analytical method that can accurately determine, for various scenarios, the threshold needed to stop spreading epidemics [13]. Among these possible scenarios are (i) immunizing people who are acquaintances of an infected individual and (ii) immunizing only those people who are acquaintances of at least two infected individuals.

Our recent results on social networks are complemented by analogous strategies for protecting other threatened networks, such as communication networks. For example, the Bar-Ilan group has already demonstrated that, in scale-free uncorrelated networks, if we immunize the neighbors of randomly-chosen sites, the critical threshold can be reduced by a factor of five [19]. This result has dramatic practical implications.

Our analytical approach is enabling us to study efficient immunization strategies in more realistic networks where, e.g., correlations, clustering effects, and geographical topology are taken into account. The immunization approach is also helping to develop methods to disintegrate targeted organizations, since by removing the nodes that are most relevant for immunity, the targeted network will collapse.

9.5 Possible Contributions of Social Network Research

(a) We are improving the tentative explanation [15] of scale-free social networks, and develop a better understanding of the range of social networks that are scale-free [16].

(b) We are developing a better understanding of the topological structures and the tomography of threatened social networks.

(c) We are developing new algorithms to improve the stability and safety of threatened networks. We are designing networks for optimal resistance to epidemics, malfunctions and attacks, and we are designing efficient and secure algorithms for organizational data flow.

(d) We are designing efficient methods for effective "immunization" that will greatly reduce spreading in threatened networks – the same mathematics describes spread of infectious agents in social networks, or "viruses" in communication networks. These methods will also help to identify weaknesses and thereby protect threatened networks.

9.6 Discussion

We are seeking to test whether concepts and methods of statistical physics such as scaling and percolation theory can be usefully applied to social networks, with special emphasis on social networks such as sexual networks and threatened networks. Many of the primary methods being used in our network research have been developed by our research group. These include the analytical percolation approach to general networks [11, 12, 13], the efficient immunization theory [19, 13], and the DFA method [26]. We also were among the first to identify scale-free networks in certain social systems and sexual networks [14, 15, 16], and we developed an approach for classifying network topologies [17].

Acknowledgments

We thank Y. Å berg, L. A. N. Amaral, C. Edling, K. Fukuda, and F. Liljeros for collaborations on which a part of this paper is based, and ONR N00014 02 1 1033 for support.

References

1. R. Albert, H. Jeong and A.-L. Barabási, "Attack and Error Tolerance of Complex Networks," *Nature* **406**, 378–382 (2000)
2. R. Albert and A.-L. Barabási, "Statistical Mechanics of Complex Networks," *Rev. Mod. Phys.* **74**, 47–97 (2002)
3. S.N. Dorogovtsev and J.F.F. Mendes, "Evolution of Networks," *Adv. in Phys.* **51**, 1079–1187 (2002)
4. R.M. Anderson, and R.M. May, *Infectious Diseases of Humans* (Oxford University Press, Oxford, 1991)
5. A.L. Lloyd and R.M. May, "How Viruses Spread among Computers and People," *Science* **292**, 1316–1317 (2001)
6. H.E. Stanley, "Scaling, Universality, and Renormalization: Three Pillars of Modern Critical Phenomena," Reviews of Modern Physics **71**, S358–S366 (1999)
7. A.-L. Barabási, *Linked: The New Science of Networks* (Perseus Publishing, Cambridge, 2002)
8. H.E. Stanley, J.S. Andrade, S. Havlin et al., "Percolation Phenomena: A Broad-Brush Introduction with Some Recent Applications to Porous Media, Liquid Water, and City Growth," Physica A **266**, 5–16 (1999)
9. A. Bunde and S. Havlin, eds., *Fractals and Disordered Systems* (Springer, New York, 1996)
10. D. ben-Avraham and S. Havlin, *Diffusion and Reactions in Fractals and Disordered Systems* (Cambridge University Press, Cambridge, 2000)

11. R. Cohen, K. Erez, D. ben-Avraham, and S. Havlin, "Resilience of the Internet to Random Breakdown," Phys. Rev. Lett. **85**, 4626 (2000)

12. R. Cohen, K. Erez, D. ben-Avraham, and S. Havlin, "Breakdown of the Internet under Intentional Attack," Phys. Rev. Lett. **86**, 3682 (2001)

13. R. Cohen, S. Havlin, and D. ben-Avraham, "Structural Properties of Scale-Free Networks" in *Handbook of Graphs and Networks: From the Genome to the Internet*, edited by S. Bornholdt and H.G. Schuster (Wiley-VCH, Berlin, in press, 2002)

14. F. Liljeros, C.R. Edling, L.A.N. Amaral, H.E. Stanley, and Y. Åberg, "The Web of Human Sexual Contacts," Nature **411**, 907–908 (2001) cond-mat/0106507

15. L.A.N. Amaral, C.R. Edling, F. Liljeros, and H.E. Stanley, "Mechanisms for the Formation of a Web of Sexual Contacts" (preprint)

16. F. Liljeros, L.A.N. Amaral, and H.E. Stanley, "Scale-Invariance in the Growth of Voluntary Organizations," Europhys. Lett. (submitted)

17. L.A.N. Amaral, A. Scala, M. Barthélémy, and H.E. Stanley, "Classes of Behavior of Small-World Networks," Proc. National Academy of Sciences **97**, 11149–11152 (2000)

18. L.A. Adamic, R.M. Lukose, A.R. Punyani, and B.A. Huberman, "Search in Power Law Networks," Phys. Rev. E **64**, 046135 (2001)

19. R. Cohen, D. ben-Avraham, and S. Havlin, "Efficient Immunization of Populations and Computers," cond-mat/0207387 (2002)

20. W.E. Leland, M.S. Taqqu, W. Willinger, and D.V. Wilson, "On the Self-Similar Nature of Ethernet Traffic (Extended Version)," IEEE/ACM Trans. Netw. **2**, 1 (1994)

21. V. Paxson and S. Floyd, "Wide Area Traffic: The failure of Poisson Modeling," IEEE/ACM Trans. Netw. **3**, 226 (1995)

22. K. Fukuda, H. Takayasu, and M. Takayasu, "Spatial and Temporal Behavior of Congestion in Internet Traffic," Fractals **7**, 23 (1999)

23. M. Takayasu, H. Takayasu, and K. Fukuda, "Dynamic Phase Transition Observed in Internet Traffic Flow," Physica A **277**, 248 (2000)

24. M. Takayasu, H. Takayasu, and K. Fukuda, "Application of Statistical Physics to Internet Traffic," Physica A **274**, 140 (1999)

25. E. Tomer, L. Safonov, and S. Havlin, "Presence of Many Stable Nonhomogeneous States in an Inertial Car-Following Model," Phys. Rev. Lett. **84**, 382 (2000)

26. C.K. Peng, S. Havlin, H.E. Stanley, and A.L. Goldberger, "Quantification of Scaling Exponents and Crossover Phenomena in Nonstationary Heartbeat Time Series" [*Proc. NATO Dynamical Disease Conference*], edited by L. Glass, Chaos **5**, 82–87 (1995)

27. C.-K. Peng, S.V. Buldyrev, S. Havlin, M. Simons, H.E. Stanley, and A.L. Goldberger, "On the Mosaic Organization of DNA Sequences," *Phys. Rev. E* **49** (1994) 1691–1695

28. S.V. Buldyrev, A.L. Goldberger, S. Havlin, R.N. Mantegna, M.E. Matsa, C.-K. Peng, M. Simons, and H.E. Stanley, "Long-Range Correlation Properties of Coding and Noncoding DNA Sequences: GenBank Analysis" Phys. Rev. E **51**, 5084–5091 (1995)

29. A. Bunde, H.J. Schellnhuber, and J. Kropp, eds., *The Science of Disasters: Climate Disruptions, Heart Attacks, and Market Crashes* (Springer-Verlag, Berlin, 2002)

30. R.B. Govindan, D. Vyushin, A. Bunde, S. Brenner, S. Havlin, and H.J. Schellnhuber, "Global Climate Models Violate Scaling of the Observed Atmospheric Variability," Phys. Rev. Lett. **89**, 028501 (2002)

31. A. Bunde, S. Havlin, J.W. Kantelhardt, T. Penzel, J.H. Peter, and K. Voigt, "Correlated and Uncorrelated Regions in Heart-Rate Fluctuations during Sleep," Phys. Rev. Lett. **85**, 3736–3739 (2000)

32. J.W. Kantelhardt, E. Koscielny-Bunde, H.H.A. Rego, S. Havlin, and A. Bunde, Physica A **294**, 441 (2001)

33. K. Hu, Z. Chen, P.Ch. Ivanov, P. Carpena, and H.E. Stanley, "Effect of Trends on Detrended Fluctuation Analysis" Phys. Rev. E **64**, 011114-1–011114-19 (2001) physics/0103018

34. Z. Chen, P.Ch. Ivanov, K. Hu, and H.E. Stanley, "Effect of Nonstationarities on Detrended Fluctuation Analysis" Phys. Rev. E **65**, 041107-1 to 041107-15 (2002). physics/0111103

35. Y. Liu, P. Gopikrishnan, P. Cizeau, M. Meyer, C.-K. Peng, and H.E. Stanley, "The statistical properties of the volatility of price fluctuations" Phys. Rev. **60**, 1390–1400 (1999)

36. P. Kumar, P.Ch. Ivanov, and H.E. Stanley, "Statistical Physics and Prime Numbers" (preprint)

37. A.F. Rozenfeld, R. Cohen, D. ben-Avraham, and S. Havlin, "Scale-Free Networks on Lattices," Phys. Rev. Lett. **89**, 218701 (2003) cond-mat/0205613

38. R. Cohen and S. Havlin, "Ultra Small World in Scale-Free Networks," Phys. Rev. Lett. **90**, 058701 (2003) cond-mat/0205476 (2002)

39. D. ben-Avraham, A. F. Rozenfeld, R. Cohen, and S. Havlin, "Geographical Embedding of Scale-Free Networks," cond-mat/0301504 (2003)

40. X. Guardiola et al., "Macro- and Microstructure of Trust networks," cond-mat/020640 (2002)

41. D. J. Watts and S.H. Strogatz, "Collective Dynamics of 'Small-World' Networks," Nature **393**, 440–442 (1998)

42. D.J. Watts, *Small Worlds: The Dynamics of Networks between Order and Randomness* (Princeton University Press, Princeton, 1999)

43. V.E. Krebs, "Mapping Networks of Terrorist Cells," Connections **24**, 43–52 (2002)

44. M.E.J. Newman, "Assortative Mixing in Networks," cond-mat/0205405 (2002)

45. R. Cohen, D. Doler, S. Havlin, T. Kalisky, O. Mohryn, and Y. Shavitt, "On the Tomography of Networks and Multicart Trees," Tech. Report TR 2002-49, Hebrew University, Israel

46. P.Ch. Ivanov, L.A. Nunes Amaral, A.L. Goldberger, S. Havlin, M.G. Rosenblum, Z. Struzik, and H.E. Stanley, "Multifractality in Human Heartbeat Dynamics," Nature **399**, 461 (1999)

47. R. Pastor-Sattoras and A. Vespignani, "Epidemic Dyanamics and Endemic States in Compex Networks," Phys. Rev. E **63**, 066117 (2001)

10 Search and Congestion in Complex Networks

Alex Arenas[1], Antonio Cabrales[2], Albert Díaz-Guilera[3], Roger Guimerà[4], and Fernando Vega-Redondo[5]

[1] Departament d'Enginyeria Informàtica, Universitat Rovira i Virgili, Tarragona
[2] Departament d'Economia i Empresa, Universitat Pompeu Fabra, Barcelona
[3] Departament de Física Fonamental, Universitat de Barcelona, Barcelona
[4] Departament d'Enginyeria Química, Universitat Rovira i Virgili, Tarragona
[5] Departament de Fonaments d'Anàlisi Econòmica, Universitat d'Alacant, Alacant

Abstract. A model of communication that is able to cope simultaneously with the problems of search and congestion is presented. We investigate the communication dynamics in model networks. Those networks consist in a regular lattice ordering plus some long-range short-cuts with a given probability. The destination nodes of the short-cuts are chosen according to some degree of preferentiality. We study then the interplay between short- and long-range links and preferentiality. We also introduce a general framework that enables a search of optimal structures. A relation between dynamical properties and topological properties of the network is found and exploited.

10.1 Introduction

In recent years, the study of static and dynamical properties of complex networks has received a lot of attention [1, 2, 3, 4, 5]. Complex networks appear in such diverse disciplines as sociology, biology, chemistry, physics or computer science. In particular, great effort has been exerted to understand the behavior of technologically based communication networks such as the Internet [6], the World Wide Web [7], or e-mail networks [8, 9, 10]. However, the study of communication processes in a wider sense is also of interest in other fields, remarkably the design of organizations [11, 12]. For instance, it is estimated that more than a half of the U.S. work force is dedicated to information processing, rather than to make or sell things in the narrow sense [11].

The pioneering work of Watts and Strogatz [1] opened a completely new field of research. Its main contribution was to show that many real-world networks have properties of random graphs and properties of regular low dimensional lattices. A model that could explain this observed behavior was missing and the proposed "small-world" model of the authors turned the interest of a large number of scientist in the statistical mechanics community in the direction of this appealing subject. Nevertheless, this simplified model gives rise to a connectivity distribution function with an exponential form, whereas many real world networks show a highly skewed degree distribution, usually with a power law tail

$$P(k) \propto k^{-\gamma} \qquad (10.1)$$

with an exponent $2 \leq \gamma \leq 3$. Barabasi and Albert [2] proposed a model where nodes and links are added to the network in such a way that the probability of the added nodes to be linked to the old nodes depend on the number of existing connections of the old node. This simple computational model can explain the power law with an exponent $\gamma = 3$.

Tools taken from statistical mechanics have been used to understand not only the topological properties of these communication networks, but also their dynamical properties. The main focus has been in the problem of searchability, although when the number of search problems that the network is trying to solve increases it raises the problem of congestion at some central nodes. It has been observed, both in real world networks [13] and in model communication networks [14, 15, 16, 17, 18], that the networks collapse when the load is above a certain threshold and the observed transition can be related to the appearance of the $1/f$ spectrum of the fluctuations in Internet flow data [19, 20].

These two problems, search and congestion, that have so far been analyzed separately in the literature can be incorporated in the same communication model. In previous works [16, 21, 18, 22] we have introduced a collection of models that captures the essential features of communication processes and are able to handle these two important issues simultaneously. In these models, agents are nodes of a network and can interchange information packets along the network links. Each agent has a certain capability that decreases as the number of packets to deliver increases. The transition from a free phase to a congested phase has been studied for different network architectures in [16, 18], whereas in [21] the cost of maintaining communication channels was considered. Finally in [22] we have attacked the problem of network optimization for fixed number of links and nodes.

This paper is organized as follows. In Sect. 2 we present well known results about search in complex networks, whereas in Sect. 3 we review recent work on network load, being considered as a betweenness centrality and hence a static characterization of the network. We present the common trends of our communication model in Sect. 4. In the next section, we show some of the exact results that have been obtained for a particular class of network, Cayley trees. Finally, in the last two sections we focus on the problem of network optimization, in the first one through a parameterized set of networks, including connectivities that can be short- or long-ranged, and different degrees of preferentiallity, and in the second one we perform an exhaustive search of optimal networks for a fixed number of nodes and links.

10.2 Search in Complex Networks

After the *discovery* of complex networks, one of the issues that has attracted a lot of attention is "search". Real complex communication networks such as the Internet or the World Wide Web are continuously changing and it is not possible to draw a *map* that allows to navigate in them. Rather, it is necessary to develop algorithms that efficiently search for the desired computers or the desired contents.

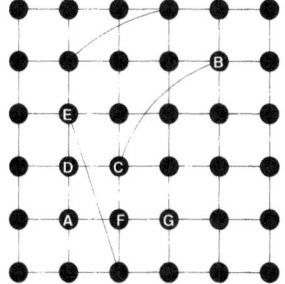

Fig. 10.1. Network topology and search in Kleinberg's scenario. Consider nodes A and B. The distance between them is $\Delta_{AB} = 6$ although the shortest path is only 3. A search process to get from A to B would proceed as follows. From A, we would jump with equal probability to D or F, since $\Delta_{DB} = \Delta_{FB} = 5$: suppose we choose F. The next jump would then be to G or C with equal probability since $\Delta_{CB} = \Delta_{GB} = 4$, although from C it is possible to jump directly to B. This is a consequence of the local knowledge of the network assumed by Kleinberg

The origin of the study of this problem is in sociology since the seminal experiment of Travers and Milgram [23]. Surprisingly, it was found that the average length of acquaintance chains was about six. This means not only that short chains exist in social networks as reported, for example, in the "small world" paper by Watts and Strogatz [1], but even more striking that these short chains can be found using local strategies, that is without knowing exactly the whole structure of the social network.

The first attempt to understand theoretically the problem of *searchability* in complex networks was provided by Kleinberg [24]. In his work, Kleinberg proposes a scenario where the network is modeled as a combination of a two-dimensional regular lattice plus a number of long-range links. The distance Δ_{ij} between two nodes i and j is defined as the number of "lattice-steps" separating them in the regular lattice, that is disregarding long-range links (see Fig. 10.1). Long range links are not established at random. Instead, when a node i establishes one of such links, it connects with higher probability with those nodes that are closer in terms of the distance Δ. In particular, the probability that the link is established with node j is

$$\Pi_{ij} \propto (\Delta_{ij})^{-r} \tag{10.2}$$

where r is a parameter.

The search algorithm proposed by Kleinberg is the following. A packet standing at one node will be sent to the neighbor of the node that is closer to the destination in terms of the distance Δ. The algorithm is local because, as shown in Fig. 10.1, the heuristics of minimizing Δ does not warrant that the packet will follow the shortest path between its current position and its destination. Therefore, the underlying two-dimensional lattice has an imprecise global informational content.

Kleinberg showed that with this essentially local scenario (with imprecise global information), short paths cannot be found in general, unless the parameter r is fixed to $r = 2$. This raised the question of why real networks are then searchable, that is, how is it possible that in real networks local strategies are able to find paths that scale as $\log N$, where N is the size of the network. Recently, Watts and coworkers have shown that with an idea similar to Kleinberg's, one can easily obtain searchable networks [25]. Their contribution consists in substituting the underlying low-dimensional lattice by an *ultra-metric* space where individuals are organized in a hierarchical fashion according to their preferences, similitudes, etc. In this case, a broad collection of networks turn out to be searchable.

Parallel to these efforts, there have been some attempts to exploit the scale free nature of some networks to design algorithms that, being local in nature, are still quite efficient [26, 27]. The idea in all these works is to profit from the scale-free nature of networks such as the Internet and bias the search towards those nodes that have a high connectivity and therefore act as hubs.

10.3 Load and Congestion in Complex Networks

When the network has to tackle several simultaneous (or parallel) search problems it raises the important issue of congestion at overburdened nodes [13, 14, 15, 16, 17]. Indeed, for a single search problem the optimal network is clearly a highly centralized star-like structure, with one or various nodes in the center and all the rest connected to them. This structure is *cheap* to assemble in terms of number of links and efficient in terms of searchability, since the average cost (number of steps) to find a given node is always bounded (2 steps), independently of the size of the system. However, the star-like structure will become inefficient when many search processes coexist in parallel in the network, due to the limitation of the central node to process all the information.

Load, independently of search, has been analyzed in different classes of networks [28, 29, 30, 31]. The load, as introduced in these works, is equivalent to the betweenness as it has been defined in social networks [32, 28]. The betweenness of a node j, β_j, is defined as the number of minimum paths connecting pairs of nodes in the network that go through node j. Among the topological properties of networks, betweenness has become one of their main characteristics. In principle the time needed for the computation of the betweenness of all vertices is of order $\mathcal{O}(MN^2)$, where N is the number of nodes and M the number of links of the network. However, Newman [28] introduced an algorithm that reduces the magnitude of the time needed for the computation by a factor of N. This definition was used to measure the social role played by scientists in some collaboration networks [28]. Later on, it was also applied to quantify model networks. Thus, in [29] different networks are constructed and their distribution of betweennesses (or loads) measured. For instance, scale-free networks with an exponent $2 < \gamma \le 3$ lead to a load distribution which is also a power law, $P(\ell) \sim \ell^{-\delta}$ with $\delta \approx 2.2$. On the other side, the load distribution of small-world networks shows a combined behavior of two Poisson-type decays. In subsequent work, the authors in [31] suggested that real-world networks should be classified in two different universality classes, according to the exponent of the power-law distribution of loads. Finally, the distribution of loads was analytically computed for scale-free trees in [30].

The works discussed in the previous paragraph consider the betweenness as a topological property of the network, since it accounts for the number of shorter-paths going through a node. However, to take into account the search algorithm and the fact that packets can perform several random steps and then go through the same node more than once we introduce an *effective* betweenness. The *effective* betweenness of node j, B_j, represents the total number of packets that would pass through j if one packet would be generated at each node at each time step with destination to any other node. The *effective* betweenness coincides with the *topological* betweenness when the nodes have complete information of the network structure and packets always follow the shortest paths between origin and destination.

10.4 A Model of Communication

The model that can handle search and congestion at the same time considers that the information is formed by discrete packets that are sent from an origin node to a destination node. Each node can store as many information packets as needed. However, the capacity of nodes to deliver information cannot be infinite. In other words, any realistic model of communication must consider that delivering, for instance, two information packets takes more time than delivering just one packet. A particular example of this would be to assume that nodes are able to deliver one (or any constant number) information packet per time step independently of their load, as happens in the communication model by Radner [11] and in simple models of computer queues [14, 15, 17], but note that many alternative situations are possible. In the present model, each node has a certain *capability* that decreases as the load of accumulated packets increases. This limitation in the capability of agents to deliver information can result in congestion of the network. Indeed, when the amount of information is too large, agents are not able to handle all the packets and some of them remain undelivered for extremely long periods of time. The maximum amount of information that a network can manage gives a measure of the quality of its organizational structure. In the study of the model, the interest is focused in both *when* the congestion occurs and *how* it occurs.

10.4.1 Description of the Model

The dynamics of the model is as follows. At each time step t, an information packet is created at every node with probability ρ. Therefore ρ is the control parameter: small values of ρ correspond to low density of packets and high values of ρ correspond to high density of packets. When a new packet is created, a destination node, different from the origin, is chosen randomly in the network. Thus, during the following time steps $t + 1, t + 2, \dots, t + T$, the packet *travels* toward its destination. Once the packet reaches the destination node, it is delivered and disappears from the network. Another interpretation is possible for this information transfer scenario. Packets can be regarded as problems that arise at a certain ratio anywhere in an organization. When one of such problems arises, it must be solved by an arbitrary agent of the network. Thus, in subsequent time steps the problem flows toward its *solution* until it is actually solved. This problem solving scenario can be considered a particularly illustrative case of the more general information transfer scenario. The problem solving interpretation suggest a model similar to Garicano's [33] in that there is task diversity and agents are specialized in solving only certain types of tasks.

The time that a packet remains in the network is related not only to the distance between the source and the target nodes, but also to the amount of packets in its path. Indeed, nodes with high loads—i.e. high quantities of accumulated packets—will need long times to deliver the packets or, in other words, it will take long times for packets to cross regions of the network that are highly congested. In particular, at each time step, all the packets move from their current position, i, to the next node in their path, j, with a probability q_{ij}. This probability q_{ij} is called the *quality of the channel* between i and j, and is defined as

$$q_{ij} = \sqrt{k_i k_j}\,, \tag{10.3}$$

where k_i represents the *capability* of agent i and, in general, changes with time. The quality of a channel is, thus, the geometric average of the capabilities of the two nodes involved, so that when one of the agents has capability 0, the channel is disabled. It is assumed that k_i depends only on the number of packets at node i, ν_i, through:

$$k_i = f(\nu_i) \tag{10.4}$$

The function $f(n)$ determines how the capability evolves when the number of packets at a given node changes. In [18] we proposed a general form although in this paper we will only show results for the case in which the number of delivered packets is constant. This particular case is consistent with simple models of computer queues [14], although the precise definition of the models may differ from ours.

The election of the functional form for the quality of the channels and the capability of the nodes is arbitrary. Regarding the first, (10.3) is plausible for situations in which an effort is needed from both agents involved in the communication process. If, on the contrary, information can be transmitted without the collaboration of the receiver, an equation of the form

$$q_{ij} = k_i\,, \tag{10.5}$$

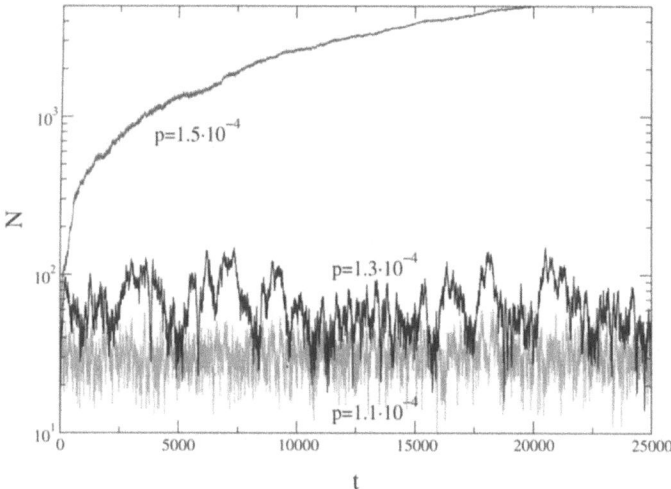

Fig. 10.2. Evolution of the total number of packets, N, as a function of time for a (5,7) Cayley tree and different values of ρ, below the critical congestion point ($\rho = 1.1 \cdot 10^{-4} < \rho_c$), above the critical congestion point ($\rho = 1.5 \cdot 10^{-4} > \rho_c$), and close to the critical congestion point ($\rho = 1.3 \cdot 10^{-4} \approx \rho_c$). Note the logarithmic scale in the Y axis

would be more adequate. Equation (10.5) will be used for analytical understanding of the problem in Sect. 10.7, whereas (10.3) is used in Sect. 10.5. Some of the most relevant features of the model, however, are not dependent on which one is used.

10.4.2 Congestion and Network Capacity

Depending on the ratio of generation of packets ρ, two different behaviors are observed. When the amount of packets is small, the network is able to deliver all the packets that are generated and, after a transient, the total load N of the network achieves a stationary state and fluctuates around a constant value. These fluctuations are indeed quite small. Conversely, when ρ is large enough the number of generated packets is larger than the number of packets that the network can manage to solve and the network enters a state of congestion. Therefore, N never reaches the stationary state but grows indefinitely in time. The transition from the *free regime*, ρ small, to the *congested regime*, ρ large, occurs for a well defined value of ρ, that will be denoted ρ_c. For values smaller than but close to ρ_c, the steady state is reached but large fluctuations arise.

The three behaviors (free, congested and close to the transition) are depicted in Fig. 10.2. For $\rho < \rho_c$, the width of the fluctuations is small, indicating short characteristic times. This means, among other thinks, that the average time required to deliver a packet to the destination is small. It also means that correlation times are short, that is, the state of the network at one time step has little influence on the state of the network only a few time steps latter. As ρ approaches ρ_c, the fluctuations are wider and one can conclude that correlations become important. In other words, as one approaches ρ_c the time needed to deliver a packet grows and the state of the network at one instant is determinant for its state many time steps later. In the congested regime, the amount of delivered packets is independent of the load and thus remains constant over time, while the number of generated packets is also constant, but larger than the amount of delivered packets. Thus, at each time step the number of accumulated packets is increased by a constant amount, and $N(t)$ grows linearly in time.

The transition from the free regime to the congested regime is therefore captured by the slope of $N(t)$ in the stationary state. When all the packets are delivered and there is no accumulation, the average slope is 0 while it is larger than 0 for $\rho > \rho_c$. We use this property to introduce an *order parameter*, η, that is able to characterize the transition from one regime to the other:

$$\eta(p) = \lim_{t \to \infty} \frac{1}{\rho S} \frac{\langle \Delta N \rangle}{\Delta t}, \tag{10.6}$$

In this equation $\Delta N = N(t + \Delta t) - N(t)$, $\langle \ldots \rangle$ indicates an average over time windows of width Δt and S is the number of nodes in the system. Essentially, the order parameter represents the ratio between undelivered and generated packets calculated at long enough times such that $\Delta N \propto \Delta t$. Thus, η is only a function of the probability of packet generation per node and time step, ρ. For $\rho > \rho_c$, the system collapses, $\langle \Delta N \rangle$ grows linearly with Δt and thus η is a function of ρ only. For $\rho < \rho_c$, $\langle \Delta N \rangle = 0$ and $\eta = 0$. Since the order parameter is continuous at ρ_c, the transition to congestion is a critical phenomenon and ρ_c is a critical point as usually defined in statistical mechanics [34].

Once the transition is characterized, the first issue that deserves attention is the location of the transition point ρ_c as a function of the parameters of the network. This transition point gives information about the capacity of a given network. Indeed, the maximum number of packets that a network can handle per time step will be $N_c = S\rho_c$. Therefore, ρ_c is a measure of the amount of information an organization is able to handle and thus of the efficiency of a given organizational structure. One reasonable problem to propose is, therefore, which is the network that maximizes ρ_c for a fixed set of available resources (agents and links).

10.5 Analytical Results for Hierarchical Lattices

As a first step we considered hierarchical networks, since they provide a zeroth order approximation to real structures, and have also been used in the economics literature to model organizations [11, 35]. In particular we are going to focus on hierarchical Cayley trees, as depicted in Fig. 10.3. Cayley trees are identified by their branching z and their number of levels m, and will be denoted (z, m) hereafter.

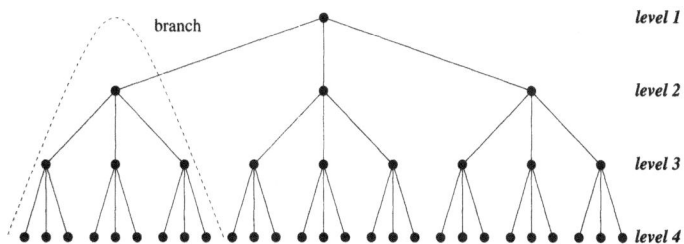

Fig. 10.3. Typical hierarchical tree structure used for simulations and calculations: in particular, it is a tree $(3, 4)$. *Dashed line:* definition of branch, as used in some of the calculations

In this case the system is regarded as hierarchical also from a knowledge point of view. It is assumed in the model that agents have complete knowledge of the structure of the network in the subbranch they root. Therefore, when an agent receives a packet, he or she can evaluate whether the destination is to be found somewhere below. If so, the packet is sent in the right direction; otherwise, the agent sends the packet to his or her supervisor. Using this simple routing algorithm, the packets travel always following the shortest path between their origin and their destination.

As happens in other problems in statistical physics [36], the particular symmetry of the hierarchical tree allows an analytical estimation of the critical point ρ_c. In particular, the approach taken here is *mean field* in the sense that fluctuations are disregarded and only average expected values are considered. By using the steady state condition that the number of packets arriving at the top node, which is the most congested one, equals the number of packets leaving it we arrive to the following inequality

$$\rho_c \geq \frac{\sqrt{z}}{\frac{z(z^{m-1}-1)^2}{z^m-1} + 1} \tag{10.7}$$

when the quality of the channels is given by (10.3). Although this expression provides an upper bound to ρ_c, (10.7) is an excellent approximation for $z \geq 3$, as shown in Fig. 10.4.

The total critical number of generated packets, $N_c = \rho_c S$, with S denoting the size of the system, can be approximated, for large enough values of z and m such that $z^{m-1} \gg 1$, by

$$N_c = \frac{z^{3/2}}{z - 1}, \tag{10.8}$$

which is independent of the number of levels in the tree. It suggests that the behavior of the top node is only affected by the total number of packets arriving from each node of the second level, which is consistent with the mean field hypothesis.

According to (10.8), the total number of packets a network can deal with, N_c, is a monotonically increasing function of z, suggesting that, given the number of agents in the organization, S, the optimal organizational structure, understood as the structure with highest capacity to handle information, is the flattest one, with $m = 2$ and $z = S - 1$.

To understand this result it is necessary to take into account the following considerations:

- We are restricting our comparison only to different hierarchical networks and in any hierarchical network, the top node will receive most of the packets. Since origins and destinations are generated with uniform independent probabilities, roughly $(z - 1)/z$ of the packets will pass through the top node.
- Still, it could seem that having small z is *slightly* better according to the previous consideration. However, it is important to note that, in the present model (in particular due to (10.3)), the loads of both the sender and the receiver are important to have a good communication quality. In a network with small z, the nodes in the second level have also a high load, while in a network with a high z the nodes in the second level are much less loaded.
- We have implicitly assumed that there is no cost for an agent to have a large amount of communication channels active.

For the order parameter, it is possible to derive an analytical expression for the simplest case where there are only two nodes that exchange packets. Since from symmetry considerations $\nu_1 = \nu_2$, the average number of packets eliminated in one time step is 2, while the number of generated packets is 2ρ. Thus $\rho_c = 1$ and with the present formulation of the model it is not possible to reach the super-critical congested regime. However, ρ can be extended to be the average number of generated packets per node at each step (instead of a probability) and in this case it can actually be as large as needed. As a result, the order parameter for the super-critical phase is $\eta = (\rho - 1)/\rho$. As observed in Fig. 10.4, the general form

$$\eta(\rho/\rho_c) = \frac{\rho/\rho_c - 1}{\rho/\rho_c} \qquad (10.9)$$

fits very accurately the behavior of the order parameter for any Cayley tree.

10.6 Optimization in Model Networks

In this section we extend previous studies about local search in model networks in two directions. First, we consider networks that, as in Kleinberg's work, are embedded in a two-dimensional space, but study the effect not only of long range random links but also of long range preferential links. Secondly and more significantly, we consider the effect of congestion when multiple searches are carried out simultaneously. As we will show, this effect has drastic consequences for optimal network design.

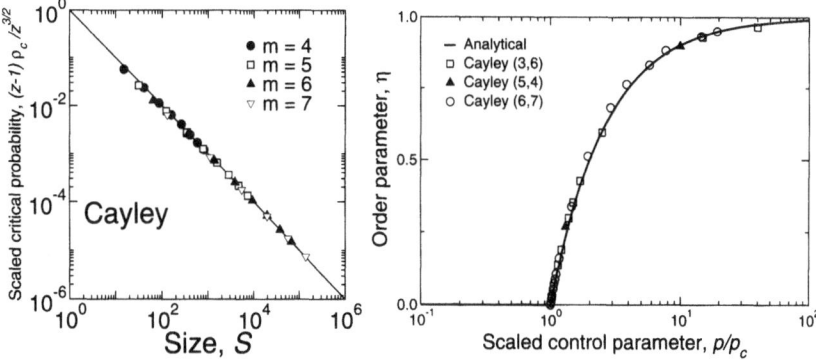

Fig. 10.4. Comparison between analytical (*lines*) and numerical (*symbols*) values obtained for hierarchical trees. *Left:* scaled critical probability (10.7). *Right:* order parameter (10.9)

10.6.1 Network Topology

The small world model [1] considered two main components: local linking with neighbors and random long range links giving rise to short average distance between nodes. The idea of Kleinberg is that local linking provides information about the social structure and can be exploited to heuristically direct the search process. Later, Barabasi and Albert showed that growth and preferential attachment play a fundamental role in the formation of many real networks [2]. Even though this model captures the correct mechanism for the emergence of highly-connected nodes, it is not likely that it captures all mechanisms responsible for the evolution of "real-world" scale-free networks. In particular, it seems plausible that in many of the networks that show scale-free behavior there is also an underlying structure as in the Watts and Strogatz model. To illustrate this idea, consider web-pages in the World Wide Web. It is plausible to assume that a page devoted to physics is more likely to be connected to another page devoted to physics than to a page devoted to sociology. That is, a set of pages devoted to physics is likely to be more inter-connected than a set including pages devoted to physics and sociology.

Therefore we consider networks with four basic components: growth, preferential attachment, local attachment and random attachment. To create the network the following algorithm is used:

1. Nodes are located in a two-dimensional square lattice without interconnecting them.
2. A node i is chosen at random.
3. We create m links starting at the selected node. With probability ϕ, the destination node is selected preferentially. With probability $1 - \phi$ the destination node is one of the nearest neighbors of the selected node. When the destination node is selected preferentially, we apply the following rule: the probability that a given destination node j is chosen is a function of its

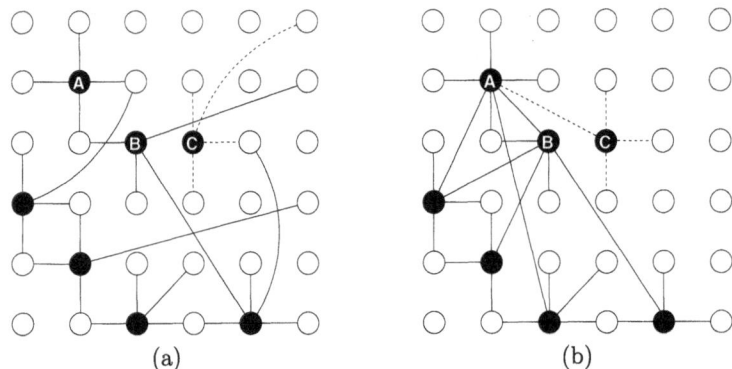

Fig. 10.5. Construction of networks with multiple linking mechanisms. In both cases $\phi = 0.25$. A random node is selected at each time step and $m = 4$ new links starting from that node are created. *Black nodes* represent nodes that have already been selected. *Dotted lines* represent the links created during the last time step in which node C was selected. In **a**, the destination of long range links is created at random ($\gamma = 0$), while in **b** they are created preferentially ($\gamma > 0$) and nodes A and B are attracting most of them

connectivity

$$\Pi_j \propto k_j^\gamma, \tag{10.10}$$

where k_j is the number of links of node j and γ is a parameter that allows to tune the network from maximum preferentiallity to no preferentiallity. Indeed, for $\gamma = 0$ the links are random and for $\gamma = 1$ we recover the BA model, that generates scale free networks in the case $\phi = 1$. For $\gamma > 1$, a few nodes tend to accumulate all the links.

4. A new node is chosen and the process is repeated from step 3, until all the nodes have been chosen once.

Figure 10.5 shows two examples of networks in the process of being created according to this algorithm. Note that in this case, the number of links is fixed and the existence of long range links implies that some local links are not present and therefore that the information contained in the two-dimensional lattice is less precise.

10.6.2 Communication Model and Search Algorithm

After the definition of the network creation algorithm, we move to the specification of the communication model and the search algorithm. For the communication model, we will use the general model presented and discussed in Sect. 10.4. As already stated, this model is general enough and considers the effect of congestion due to limitation of ability of nodes to handle information.

In comparison with hierarchical networks, there is only one ingredient of the communication model that needs to be reformulated. In the hierarchical version of the model, when a node receives a packet, it decides to send it downwards in the right direction if the solution is there, or upward to the agent overseeing her otherwise. This simple *routing algorithm* arises from the fact that we implicitly assume that the hierarchy is not only a communicational hierarchy, but also a knowledge hierarchy, where nodes know perfectly the structure of the network *below* them. In a complex network, this informational content of the hierarchy is lost. Here we will use Kleinberg's approach [24]. When an agent receives a packet, she knows the coordinates in the underlying two-dimensional space of its destination. Therefore, she forwards the packet to the neighbor that is *closer* to the destination according to the lattice distance Δ defined in Sect. 10.2, provided that the packet has not visited that node previously[6]. Note, however, that distance refers to the two-dimensional space, but not necessarily to the topology of the complex network and, as in Kleinberg's work, there might be shortcuts in directions that increase Δ. Moreover, here long range links *replace* short range links and are not simply added to short range links. Therefore it is possible that following the direction of minimization of Δ the packet arrives to a dead end and has to go back.

Considering this algorithm, it is interesting that the three mechanisms to establish links (local, random and preferential) are somehow complementary. A completely regular lattice (all links are local) contains a lot of information since all the agents efficiently send their packets in the best possible direction. However, the average path length is extremely high in this networks and therefore the number of packets that are flowing in the network at a given time is also very high. The addition of random links can reduce dramatically the average path length, as in small world networks. However, if the number of random links is very high, then the number of local links is small and thus sending the packet to the node closer to the destination is probably quite inefficient (since it is possible that, even if it is very close in the underlying two-dimensional space, there is no short path in the actual topology of the network). Finally, preferential links seem to solve both problems. They obviously solve the long average path length problem but, in addition, the loss of information is not large, since the highly connected that actually concentrate this information. The star configuration is an extreme example of this: although there are no local links, the central node is capable of sending all the packets in the right directions. However, when the amount of information to handle is big, preferential links are especially inadequate because highly connected nodes act as centers of congestion. Therefore, optimal structures should be networks where all the mechanisms coexist: complex networks.

10.6.3 Results

We simulate the behavior of the communication model in networks built according to the algorithm presented in Sect. 10.4.1. First, a value of the probability

---

[6] Packets are sent to previously visited nodes only if it is strictly necessary. This *memory* restriction avoids packets getting trapped in loops

of packet generation per node and time step, ρ, is fixed. For that particular value, we compare the performance of different networks: networks with different preferentiallity, from random ($\gamma = 0$) to maximum centralization ($\gamma \gg 1$), and with different fraction of long range links, from pure regular lattices with no long range links ($\phi = 0$) to networks with no local component ($\phi = 1$). For each collection of the parameters ρ, γ, and ϕ, the network load, $\overline{N}$, is calculated and averaged over a certain time window and over 100 realizations of the network, so that fluctuations due to particular simulations of the packet generation and of the network creation are minimized. As in the economics literature, the objective is to minimize the average time τ for a packet to go from the origin to the destination.

According to Little's Law of queuing theory [37], the characteristic time is proportional to the average total load, $\overline{N}$, of the network:

$$\frac{\overline{N}}{\tau} = \rho S \Rightarrow \tau = \frac{\overline{N}}{\rho S} \qquad (10.11)$$

where ρ is the probability of packet generation for each node at each time step. Thus, minimizing the average cost of a search is equivalent to minimizing the total load $\overline{N}$ of the network.

The main results are shown in Fig. 10.6. Consider first the behavior of the networks at low values of ρ. Figure 10.6a shows the load of the network for $\rho = 0.01$ as a function of the fraction of long range links, ϕ, both when they are random $\gamma = 0$ and when they are extremely preferential $\gamma = 6$. In the last case, long range links are established only with the most connected node. In this case of small ρ, centralization is not a big problem because congestion effects are still not important. Therefore, preferential links are, in general, better than random long range links. In the case of preferential links, it is interesting to understand the behavior of the curve $\overline{N}(\phi)$. For $\phi = 0$ the network is a two-dimensional regular lattice and then the average distance between nodes is large. As some long range links are introduced, the average path length decreases as in the Watts-Strogatz model [1], and therefore the load of the network is smaller because packets reach their destination faster. However, the addition of long range links implies the lack of local links and when ϕ is further increased, the heuristic of minimizing the lattice distance Δ becomes worse and worse. This fact explains that for $\phi \approx 0.15$ (the network is similar to the one depicted in Fig. 10.6d) the load has a local minimum that arises due to the trade-off between the two effects of introducing long range preferential links: shortening of the distances that tends to decrease $\overline{N}$ and destruction of the lattice structure that tends to decrease the utility of the heuristic search and then to increase $\overline{N}$. If ϕ is further increased, one node tends to concentrate all the links and for $\phi = 1$ (Fig. 10.6e) the network is strictly a star with one central node and the rest connected to it. In this completely centralized situation, the lack of two-dimensional lattice is not important because the packets will be sent to the central node and from there directly to the destination. Since for small ρ congestion is not an issue,

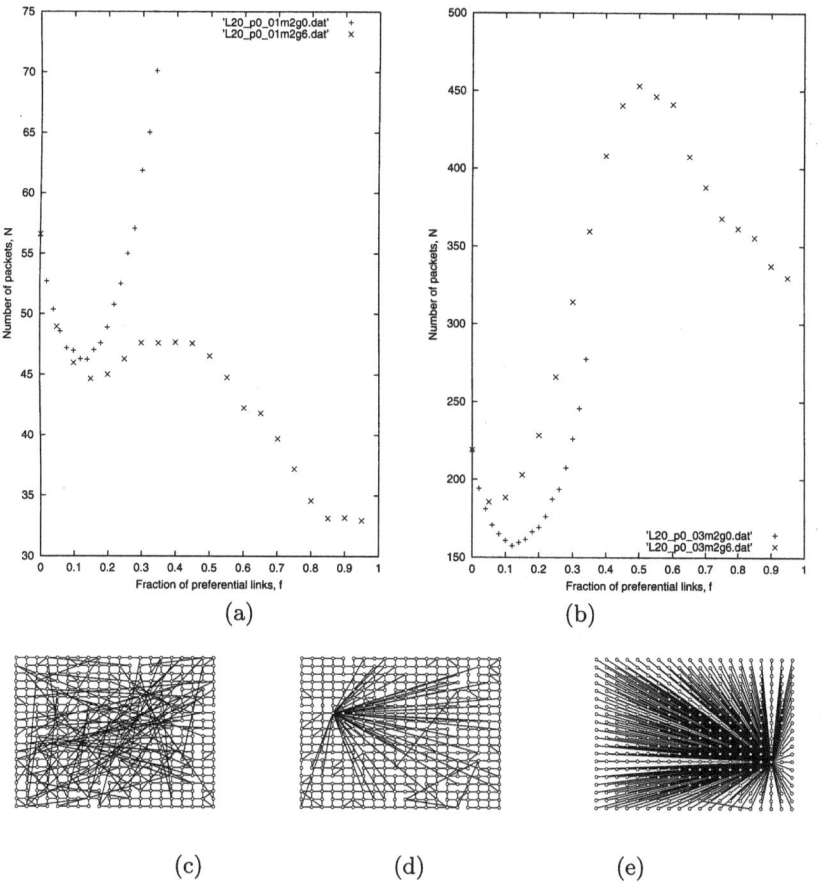

Fig. 10.6. **a** and **b** Average number of packets flowing in the network as a function of the fraction of preferential links: **a** $\rho = 0.01$ and **b** $\rho = 0.03$. Symbol (+) corresponds to $\gamma = 0$ (random links) and symbol (×) corresponds to $\gamma = 6$ (extremely focused links). Figures **c**, **d** and **e** show the typical shape of complex networks with particularly efficient configurations: **c** $\gamma = 0$ and $\phi = 0.12$; **d** $\gamma = 6$ and $\phi = 0.07$; and **e** $\gamma = 6$ and $\phi = 1.0$

this structure turns out to be even better than the locally optimal structure with $\phi \approx 0.15$.

The situation is different when considering higher values of the probability of packet generation (Fig. 10.6b displays the the results for $\rho = 0.03$). Regarding preferential linking, the two locally optimal structures with $\phi = 0.7$ and $\phi = 1$ (Figs. 10.6d and 10.6e respectively) persist. However, in this situation and due to congestion considerations the first is better than the second. Thus, at some intermediate value of $0.01 < \rho < 0.03$, there is a transition such that the optimal structure changes from being the star configuration to being the *mixed* configuration with local as well as preferential links. Significantly, this transition is sharp, meaning that there is not a continuous pass from the star to the mixed.

Beyond the behavior of networks built with preferential long range links, it is worth noting that when the effect of the congestion is important (Fig. 10.6b), the structure depicted in Fig. 10.6c, where the long range links are actually thrown at random, becomes better than the structure in 10.6.d. In other words, the optimal network is, in this case, a completely decentralized small world network *a la* Watts-Strogatz.

10.7 Optimization in a General Framework

In the previous section we have compared the behavior of networks which have been built following different rules (nearest neighbor linking, preferential attachment, etc.). The main reason for focusing on a particular set of networks is that it is very costly to compare the performance of two networks: it is necessary to run a simulation, wait for the stationary state and calculate the average load of the network. Specially, close to the critical point the time needed to reach the stationary state diverges. In [22] we presented a formalism that is able to cope with search and congestion simultaneously, allowing the determination of optimal topologies. This formalism avoids the problem of simulating the dynamics of the communication process and provides a general scenario applicable to any communication process.

Let us focus on a single information packet at node i whose destination is node k. The probability for the packet to go from i to a new node j in its next movement is p_{ij}^k. In particular, $p_{kj}^k = 0 \; \forall j$ so that the packet is *removed* as soon as it arrives to its destination. This formulation is completely general, and the precise form of p_{ij}^k will depend on the search algorithm and on the connectivity matrix of the network. In particular, when the search is Markovian, p_{ij}^k does not depend on previous positions of the packet. In this case, the probability of going from i to j in n steps is given by

$$P_{ij}^k(n) = \sum_{l_1,l_2,\dots,l_{n-1}} p_{il_1}^k p_{l_1 l_2}^k \cdots p_{l_{n-1}j}^k. \tag{10.12}$$

This definition allows us to compute the average number of times, b_{ij}^k, that a packet generated at i and with destination at k passes through j.

$$b^k = \sum_{n=1}^{\infty} P^k(n) = \sum_{n=1}^{\infty} \left(p^k\right)^n = (I - p^k)^{-1} p^k. \tag{10.13}$$

and the effective betweenness of node j, B_j, is then defined as the sum over all possible origins and destinations of the packets,

$$B_j = \sum_{i,k} b_{ij}^k. \tag{10.14}$$

When the search algorithm is able to find the minimum paths between nodes, the effective betweenness will coincide with the topological betweenness, β_j, as usually defined [32, 28].

Once, these quantities have been defined, we focus on the load of the network, $N(t)$, which is the number of floating packets. These floating packets are stored in the nodes that act as queues. In a general scenario where packets are generated at random and independently at each node with a probability ρ, the arrival of packets to a given node j is a Poisson process. In the original model presented in Sect. 10.4 we assumed that the quality of the channels depend on both the sender and the receiver nodes; if one assumes that it only depends on the receiver node then the delivery of packets is also a Poisson process. In this simple picture, the queues are called M/M/1 in the computer science literature and the average load of the network is [37, 22]

$$\overline{N} = \sum_{j=1}^{S} \frac{\frac{\rho B_j}{S-1}}{1 - \frac{\rho B_j}{S-1}}. \tag{10.15}$$

There are two interesting limiting cases of equation (10.15). When ρ is very small, taking into account that the sum of betweennesses is proportional to the average distance, one obtains that the load is proportional to the average effective distance. On the other hand, when ρ approaches ρ_c most of the load of the network comes from the most congested node, and therefore

$$\overline{N} \approx \frac{1}{1 - \frac{\rho B^*}{S-1}} \qquad \rho \to \rho_c, \tag{10.16}$$

where B^* is the effective betweenness of the most central node. The last results suggest the following interesting problem: to minimize the load of a network it is necessary to minimize the effective distance between nodes if the amount of packets is small, but it is necessary to minimize the largest effective betweenness of the network if the amount of packets is large. The first is accomplished by a *star-like* network, that is, a network with one central node and all the others connected to it. The second, however, is accomplished by a very decentralized network in which all the nodes support a similar load. This behavior is similar to any system of queues provided that the communication depends only on the sender.

It is worth noting that there are only two assumptions in the calculations above. The first one has already been mentioned: the movement of the packets needs to be Markovian to define the jump probability matrices p^k. Although this is not strictly true in real communication networks—where packets are not usually allowed to go through a given node more than once—it can be seen as a first approximation [14, 16, 17]. The second assumption is that the jump probabilities p_{ij}^k do not depend on the congestion state of the network, although communication protocols sometimes try to avoid congested regions, and then $B_j = B_j(\rho)$. However, all the derivations above will still be true in a number of general situations, including situations in which the paths that the packets follow are unique, in which the routing tables are fixed, or situations in which the structure of the network is very homogeneous and thus the congestion of all the nodes is similar. Compared to situations in which packets avoid congested regions, it correspond to the worst case scenario and thus provide bounds to more realistic scenarios in which the search algorithm interactively avoids congestion.

Equation (10.15) relates a dynamical variable, the load, with the topological properties of the network and the properties of the algorithm. So we have converted a dynamical communication problem into a topological problem. Hence, the dynamical optimization procedure of finding the structure that gives the minimum load is reduced to a topological optimization procedure where the network is characterized completely by its effective betweenness distribution. In [22] we considered the problem of finding optimal structures for a purely local search, using a generalized simulated annealing (GSA) procedure, as described in [38, 39]. On the one side, we have found that for $\rho \to 0$ the optimal network has a star-like centralized structure as expected, which corresponds to the minimization of the average effective distance between nodes. On the other extreme, for high values of ρ, the optimal structure has to minimize the maximum betweenness of the network; this is accomplished by creating a homogeneous network where all the nodes have essentially the same degree, betweenness, etc. One could expect that the transition centralized-decentralized occurs progressively. Surprisingly, the results of the optimization process reveal a completely different scenario. According to simulations, star-like configurations are optimal for $\rho < \rho^*$; at this point, the homogeneous networks that minimize B^* become optimal. Therefore there are only two type of structures that can be optimal for a local search process: star-like networks for $\rho < \rho^*$ and homogeneous networks for $\rho > \rho^*$.

Beyond the existence of both centralized and decentralized optimal networks, it is significant that the transition from one sort of networks to the other is abrupt, meaning that there are no intermediate optimal structures between total centralization and total decentralization. As already mentioned, this property is shared by the model networks in the previous section. Our explanation of this fact is the following. Since we are considering (in both the present and the last sections) local knowledge of the network topology, centered star-like configurations are extremely efficient in searching destinations and thus minimizing the effective distance between nodes. This explains that stars are optimal for a wide range of values of ρ, until the central node (or nodes) becomes congested. At this point, structures similar to stars will have the same problem and will be much worse regarding search; at this point, the only alternative is something completely decentralized, where the absence of congestion can compensate the dramatic increase in the effective distance between nodes. If this explanation is correct, one should be able to obtain a smooth transition from centralization to decentralization by considering global knowledge of the network, in such a way that the average effective distance (that in this case coincides with the average path length) is not much larger in an arbitrary network than in the star. Alt-

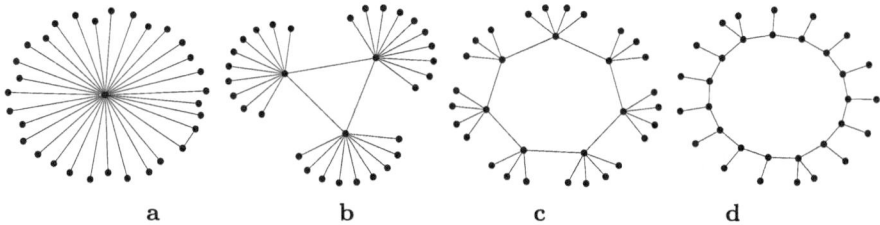

Fig. 10.7. Optimal topologies for networks with $S = 32$ nodes, $L = 32$ links and global knowledge. **a** $\rho = 0.010$. **b** $\rho = 0.020$. **c** $\rho = 0.050$. **d** $\rho = 0.080$. In this case of global knowledge, the transition from centralization to decentralization seems smooth

hough we do not have extensive simulations in this case, Fig. 10.7 shows that there is some evidence to think that this is indeed the case.

10.8 Summary

We have presented some results concerning search and congestion in networks. By defining a communication model we have been able to cope with the problems of search and congestion simultaneously. For a hierarchical lattice some analytical results are found, by exploiting the symmetry properties of the network. For complex networks, this is not the case, and computational optimization to look for the best structures is required. On the one hand, for model networks where short-range, long-range, random and preferential connections are mixed we find that network that perform well for very low load become easily congested when the load is increased. On the other hand, when searching for optimal structures in a general scenario there is a clear transition from star-like centralized structures to homogeneous decentralized ones.

Acknowledgments

This work has been supported by DGES of the Spanish Government, Grants No. PPQ2001-1519, No. BFM2000-0626, No. BEC2000-1029, and No. BEC2001-0980, and EC-FET Open Project No. IST-2001-33555.

References

1. D. Watts, S. Strogatz: Nature **393**, 440 (1998)
2. A.-L. Barabasi, R. Albert: Science **286**, 509 (1999)
3. L.A.N. Amaral, A. Scala, M. Barthelemy, H.E. Stanley: Proc. Nat. Acad. Sci. USA **97**, 11149 (2000)
4. R. Albert, A.-L. Barabasi: Rev. Mod. Phys. **74**, 47(2002)
5. S. Dorogovtsev, J.F.F. Mendes: Adv. Phys. **51**, 1079 (2002)

194   A. Arenas et al.

6. M. Faloutsos, P. Faloutsos, C. Faloutsos: Comp. Comm. Rev. **29**, 251 (1999)
7. R. Albert, H. Jeong, A.-L. Barabasi: Nature **401**, 130 (1999)
8. M.E.J. Newman: Phys Rev E **66**, 035101 (2002)
9. H. Ebel, L.-I. Mielsch, S. Bornholdt: Phys Rev E **66**, 035103 (2002)
10. R. Guimera, L. Danon, A. Diaz-Guilera, F. Giralt, A. Arenas: cond-mat/0211498.
11. R. Radner: Econometrica **61**, 1109 (1993)
12. S. DeCanio, W. Watkins: J. Econ. Behavior and Organization **36**, 275 (1998)
13. V. Jacobson: 'Congestion avoidance and control'. In: *Proceedings of SIGCOMM '88* (ACM, Standford, CA 1988)
14. T. Ohira, R. Sawatari: Phys. Rev. E **58**, 193 (1998)
15. A. Tretyakov, H. Takayasu, M. Takayasu: Physica A **253**, 315 (1998)
16. A. Arenas, A. Diaz-Guilera, R. Guimera: Phys. Rev. Lett. **86**, 3196 (2001)
17. R. Sole, S. Valverde: Physica A **289**, 595(2001)
18. R. Guimera, A. Arenas, A. Diaz-Guilera, F. Giralt: Phys. Rev. E **66**, 026704 (2002)
19. I. Csabai: J. Phys. A: Math. Gen. **27**, L417 (1994)
20. M. Takayasu, H. Takayasu, T. Sato: Physica A **233**, 824(1996)
21. R. Guimera, A. Arenas, A. Diaz-Guilera: Physica A **299**, 247 (2001)
22. R. Guimera, A. Diaz-Guilera, F. Vega-Redondo, A. Cabrales, A. Arenas: Phys. Rev. Lett. **89**, 248701 (2002)
23. J. Travers, S. Milgram, Sociometry **32**, 425 (1969)
24. J. Kleinberg: Nature **406**, 845 (2000)
25. D.J. Watts, P.S. Dodds, M.E.J. Newman: Science **296**, 1302 (2002)
26. L.A. Adamic, R.M. Lukose, A.R. Puniyani, B.A. Huberman: Phys. Rev. E **64**, 046135 (2001)
27. B. Tadic: Eur. Phys. J. B **23**, 221 (2001)
28. M.E.J. Newman: Phys. Rev. E **64**, 016132 (2001)
29. K.-I. Goh, B. Kahng, D. Kim: Phys. Rev. Lett. **87**, 278701 (2001)
30. G. Szabó, M. Alava, J. Kertész: Phys. Rev. E **66**, 026101 (2002)
31. K.-I. Goh, E. Oh, H. Jeong, B. Kahng, D. Kim: Proc. Nat. Acad. Sci. USA **99**, 12583 (2002)
32. L. C. Freeman, Sociometry **40**, 35 (1977)
33. L. Garicano: J. Political Economy **108**, 874 (2000)
34. H.E. Stanley: *Introduction to phase transitions and critical phenomena* (Oxford University Press, Oxford 1987)
35. P. Bolton, M. Dewatripont: Quart. J. Economics **109**, 809 (1994)
36. D. Stauffer, A. Aharony: *Introduction to percolation theory*, 2nd edn. (Taylor and Francis, 1992)
37. O. Allen: *Probability, statistics and queueing theory with computer science application*, 2nd edn. (Academic Press, New York 1990)
38. C. Tsallis, D.A. Stariolo: In *Annual Review of Computational Physics II*, ed. by D. Stauffer (World Scientific, Singapore 1994)
39. T.J.P. Penna: Phys. Rev. E **51**, R1 (1995)

11 Membrane Clusters of Ion Channels: Size Effects for Stochastic Resonance

Gerhard Schmid, Igor Goychuk, and Peter Hänggi

Universität Augsburg, Institut für Physik, D-86135 Augsburg, Germany

Abstract. By use of a stochastic generalization of the Hodgkin-Huxley model we investigate the phenomenon of *Stochastic Resonance* (SR) for a distribution of ion channels within a cluster of variable size. In the presence of a periodic stimulus we demonstrate *intrinsic SR* vs. decreasing patch size, or, put differently, vs. increasing internal noise strength. SR with external noise occurs only for large cluster sizes which possess suboptimal internal noise levels. In particular, SR in biology thus seemingly is rooted in the collective properties of optimally selected ion channel assemblies. Moreover, upon investigating the signal-to-noise ratio (SNR) for sub-threshold sinusoidal driving vs. driving frequency we encounter also a stochastic resonance behavior which reflects the existence of a random internal limit cycle. The occurrence of *intrinsic* SR in a combination with the conventional frequency resonance may be of importance for the frequency tuning in biological signal processing.

11.1 Introduction

Much attention is presently given to the behavior of complex networks with the particular focus being on so termed scale-free networks, which are believed to present many complex phenomena in nature [1, 2, 3, 4, 5, 6, 7]. Such networks naturally also occur in biological settings. In this spirit we focus here on the constructive role of noise on voltage gated, globally connected assemblies of ion channels. If the distribution of such ion channels consists of at least two types, excitable behavior becomes possible which in turn rules the transduction of biological information. The transduction of signals in presence of ambient, internal noise then likely makes use of a cooperative behavior between nonlinearity and noise, known under the label of *Stochastic Resonance* [8].

During the last decade, the effect of *Stochastic Resonance* (SR) – a cooperative phenomenon wherein the addition of external noise improves the detection and transduction of signals in nonlinear systems (for comprehensive surveys and relevant further references, see in [8, 9]) – has been studied experimentally and theoretically in various biological systems [10, 11, 12, 13, 14]. For example, SR has been experimentally demonstrated within the mechanoreceptive system in crayfish [10], in the cricket cercal sensory system [11], for human tactile sensation [12], visual perception [13], and response behavior of the arterial baroreflex system of humans [14]. The importance of this SR-phenomenon for sensory biology is by now well established; yet, it is presently not known to which minimal level of the biological organization the stochastic resonance effect can ultimately be traced down. Presumably, SR has its origin in the stochastic properties of the ion channel clusters located in a receptor cell membrane. Indeed, for an artificial model system Bezrukov and Vodyanoy have demonstrated experimentally that a finite ensemble of the alamethicin ion channels does exhibit stochastic resonance [15]. This in turn provokes the question whether a *single* ion channel is able to exhibit SR, or whether stochastic resonance is the result of a *collective* response from a finite assembly of channels.

Stochastic resonance in single, biological potassium ion channels has also been investigated both theoretically [16] and experimentally [17]. Thus far, the experimental work did not convincingly reveal SR in single voltage-sensitive ion channels versus the varying temperature. Nevertheless, the SR phenomenon versus the *externally* added noise can occur in single ion channels if only the parameters are within a regime where the channel is predominantly dwelled in the closed state, as demonstrated within a theoretical modeling for a Shaker potassium channel [16]. The manifestation of SR on the *single*-molecular level, is not only of academic interest, but is also relevant for potential nano-technological applications, such as the design of single-molecular bio-sensors. The origin and biological relevance of SR in single ion channels, however, remains still open.

Indeed, biological SR is a manifestation of *collective* properties of large assemblies of ion channels of different sorts. To display the phenomenon of excitability these assemblies must contain an assemblage of ion channels of at least two different sorts – such as, *e.g.*, potassium and sodium channels. The corresponding mean-field model has been put forward by Hodgkin and Huxley as early as in 1952 [18] by neglecting the intrinsic fluctuations which originate from the stochastic opening and closing of channels. SR due to *external* noise in this primary model and related models of excitable dynamics has extensively been addressed [19, 20]. A challenge though still remains: does *internal* noise play a constructive role for SR? Internal noise is produced by fluctuation of the number of open channels within the assembly, and diminishes with increasing number of channels. For a large, macroscopic number of channels this noise becomes negligible. Under the realistic biological conditions, however, it may play an important role [21].

11.2 The Hodgkin-Huxley Model

Our starting point is the well-established model of Hodgkin and Huxley [18]. The membrane patch of area S is considered as an electrical capacitor possessing the specific area capacitance C. The membrane separates two ionic bath solutions (which in vivo correspond to the interior and the exterior of the excitable cell) with different concentrations of the ions of different sorts, mainly potassium, K^+, sodium, Na^+, and chloride, Cl^- ions. The macroscopic concentration differences are kept approximately constant. In the cell, this task is accomplished by the ATP-driven ionic pumps. Furthermore, the ionic baths are on the average electrically neutral. However, due to the different ionic concentrations on the opposite sides of the semi-permeable membrane, the membrane becomes charged. As a consequence, an equilibrium transmembrane electrical potential difference emerges. The lipid membrane creates an almost impenetrable barrier for the ions. However, they can flow across the membrane through special ion selective pores created by specialized membrane proteins – the ion channels [22]. The specific potassium, I_K, and sodium, I_{Na}, ion currents through the open ion channels are approximately proportional to the differences of the transmembrane potential V and the specific (for the particular sort of ions) equilibrium potentials, E_K and E_{Na}, respectively. The *stochastically* averaged, mean conductances, $G_{Na}(m,h)$ and $G_K(n)$, are, however, strongly nonlinear functions of V. This nonlinearity emerges due to the gating dynamics (see below). There exists also the leakage current I_L, mainly due to the chloride ions. If the membrane is driven by the external current $I_{ext}(t)$, the sum of the specific ion currents and the capacitive current, I_C, must be equal to $I_{ext}(t)$ as a consequence of the charge conservation. Therefore, the equation for the transmembrane potential $V(t)$ reads

$$C\frac{d}{dt}V + G_K(n)\ (V - E_K) + G_{Na}(m,h)\ (V - E_{Na})$$

$$+ G_L\ (V - E_L) = I_{ext}(t)\,. \quad (11.1)$$

For a squid giant axon, the parameters in eq. (11.1) are $E_{Na} = 50\,\text{mV}$, $E_K = -77\,\text{mV}$, $E_L = -54.4\,\text{mV}$, and $C = 1\,\mu\text{F/cm}^2$. Furthermore, the leakage conductance is assumed to be constant, $G_L = 0.3\,\text{mS/cm}^2$. On the contrary, the sodium and potassium conductances are controlled by the voltage-dependent gating dynamics of single ion channels and are proportional to their respective numbers. These latter assumptions have been fully confirmed in the single-channel recordings by Neher, Sakmann and colleagues which indeed have proven that ion channels undergo the opening-closing stochastic gating dynamics [23]. In the Hodgkin-Huxley model, the opening of the potassium ion channel is governed by four identical activation gates characterized by the opening probability n. The channel is open when all four gates are open. In the case of sodium channel, the dynamics is governed by the three independent, identical fast activation gates (m) and an additional slow, so-termed inactivation gate (h). The independence of the gates implies that the probability $P_{K,Na}$ of the occurrence of the conducting conformation is $P_K = n^4$ for a potassium channel and $P_{Na} = m^3 h$ for a sodium channel, respectively. In the mean-field description, the macroscopic potassium and sodium conductances thus read:

$$G_K(n) = g_K^{max}\ n^4, \quad G_{Na}(m,h) = g_{Na}^{max}\ m^3 h\,, \quad (11.2)$$

where $g_K^{max} = 36\,\text{mS/cm}^2$ and $g_{Na}^{max} = 120\,\text{mS/cm}^2$ denote the maximal conductances (when all channels are open). The two-state, open–closing dynamics of the gates is given by the voltage dependent opening and closing rates $\alpha_x(V)$ and $\beta_x(V)$ ($x = m, h, n$), i.e.

$$\alpha_m(V) = \frac{0.1(V + 40)}{1 - \exp[-(V + 40)/10]}, \tag{11.3a}$$

$$\beta_m(V) = 4 \ \exp[-(V + 65)/18], \tag{11.3b}$$

$$\alpha_h(V) = 0.07 \ \exp[-(V + 65)/20], \tag{11.3c}$$

$$\beta_h(V) = \{1 + \exp[-(V + 35)/10]\}^{-1}, \tag{11.3d}$$

$$\alpha_n(V) = \frac{0.01 \ (V + 55)}{1 - \exp[-(V + 55)/10]}, \tag{11.3e}$$

$$\beta_n(V) = 0.125 \ \exp[-(V + 65)/80]. \tag{11.3f}$$

Hence, the dynamics of the opening probabilities for the gates are given by:

$$\dot{x} = \alpha_x(V) \ (1 - x) - \beta_x(V) \ x, \quad x = m, h, n. \tag{11.4}$$

The voltage equation (11.1), (11.2) and the rate equations of the gating dynamics (11.3), (11.4) define the original, purely deterministic Hodgkin–Huxley model [18] for the squid giant axon.

The rate constants in (11.3) are given in ms^{-1} and the voltage in mV. These nonlinear Hodgkin-Huxley equations (11.1)–(11.3) present a cornerstone model in neurophysiology. Within the same line of reasoning this model can be generalized to a mixture of different ion channels with various gating properties [24, 25].

The dynamics of the Hodgkin-Huxley model exhibits a complex, rich behavior which sensitively depends on the model parameters. For the squid giant axon parameters, the corresponding dynamics possesses a single fixed point and therefore does not exhibit a spiking activity in the absence of external stimulus, $I_{ext}(t) = 0$. However, if a constant stimulus, $I_{ext}(t) = I_0$, is applied, the fixed point loses its stability with increasing strength I_0 upon $I_0 \geq I_1 \approx 9.763 \ \mu A/cm^2$. For such a super-threshold current strength, the membrane exhibits a periodic spiking activity which reflects the presence of a stable limit cycle, see Fig. 11.1a. Upon decreasing the driving current strength, the spiking dynamics still persists below the threshold for excitation, i.e. also for $I < I_1$, until the diminishing current reaches the sub-critical value $I_2 \approx 6.26 \ \mu A/cm^2$. Below this value, the limit cycle loses stability and the spiking activity vanishes. In conclusion, for $I_2 < I_0 < I_1$ both the stable fixed point and the stable limit cycle can indeed coexist. This feature is thus responsible for the hysteresis behavior in the spiking behavior versus the varying driving current strength, cf. Fig. 11.1a.

Next we focus on a periodic sinusoidal driving,

$$I_{ext}(t) = A \ \sin(\Omega t), \tag{11.5}$$

with amplitude strength A and angular driving frequency Ω, see also in [26]. The corresponding dynamics becomes now even richer. In this case of periodic driving, the firing threshold A_{th} becomes frequency-dependent. The corresponding complexity for the phase diagram is depicted in Fig. 11.1b. In Fig. 11.1b, the border line $A_{th}(\Omega)$ which separates the regime of no spiking from the regime

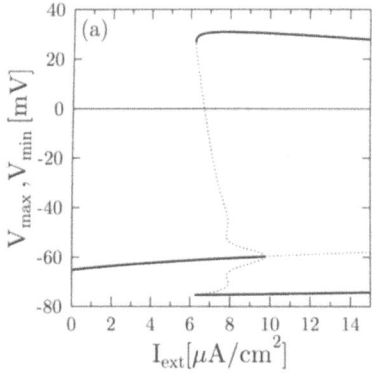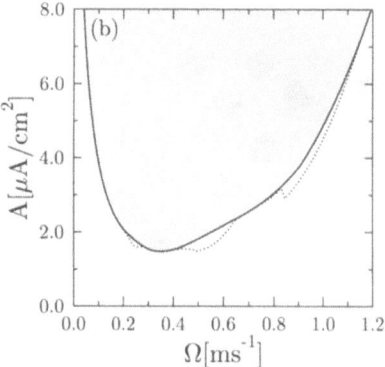

Fig. 11.1. The bifurcation diagram for the emergence of spiking behavior for the deterministic Hodgkin-Huxley model. The equilibrium voltage (fixed point) and the minimal and maximal voltage amplitudes, respectively, of the limit cycle oscillations are plotted in part **a** against the constant driving current strength. There exists a hysteretic behavior for the range $6.26 < I_{ext} < 9.763$ µA/cm^2 where the stable limit cycle exhibiting firing events and the stable fixed point coexist. **b** In the regime of periodic sinusoidal driving the situation is rather complex as well: a phase diagram is plotted as a function of the driving frequency Ω and the corresponding threshold amplitude A_{th} for firing. The solid line separates the phase with firing events (*grey region*) from the phase without spike occurrences (*white region*) for the case when the amplitude is correspondingly increased from zero. Upon starting from the regime with spiking the *dotted line* gives the phase separation line when the transition into the no-firing regime occurs. This reflects a typical hysteretic character

with an assured spiking behavior exhibits a hysteresis-like character. The solid line in Fig. 11.1b marks the transition to spiking when the driving strength is successively increased from zero driving strength. The occurrence of a minimum in this figure is quite remarkable: It implies that the system possesses an internal resonance frequency. This feature can be used for signal processing [20, 26].

11.3 Stochastic Version of the Hodgkin-Huxley Model

It has been suspected since the time of Hodgkin and Huxley, and known with certainty since the first single-channel recordings of Neher, Sakmann and colleagues, that voltage-gated ion channels are stochastic devices [23]. An essential drawback of the Hodgkin-Huxley model, however, is that it operates with the *average* number of open channels, thereby disregarding the corresponding number fluctuations (or, the so-called *channel noise* [23, 21]). These fluctuations, i.e. their strength, scale inversely proportional to the number of ion channels, see below. Thus, the original Hodgkin-Huxley model can be valid, strictly speaking, only within the limit of very large system size. We emphasize, however, that the size of an excitable membrane patch within a neuron is typically finite.

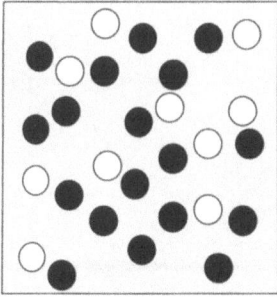

Fig. 11.2. Sketch of a membrane patch with potassium (*white filled circles*) and sodium *(black filled circles)* ion channels. The *grey background* indicates the leakage caused by additional non-voltage dependent channels. The ion channels interact only globally, through the membrane voltage

In a small spherical neuron about 10 μm in diameter, the membrane area is about 300 μm^2. Therefore, for a modest channel density of $\rho = 30$ μm^{-2} there should be about $N = 9000$ ion channels. The ratio of the standard deviation to the mean conductance of the whole ensemble of identical channels, which is $\delta G = \sqrt{(1 - P_o)/(P_o N)}$, where P_o the stationary opening probability of a channel (see, e.g., in [21]), scales as $\delta G \propto 1/\sqrt{N}$ with the number N of ion channels. Therefore, the conductance fluctuations may become appreciable for a small cell. These fluctuations can also play a functional role [21]. Besides, the spatial distribution of channels in receptor cell membranes is highly inhomogeneous and occurs in the form of clusters (see, e.g., an example in [27]) which are electrically coupled through the electrically passive pieces of membrane. As a consequence, the role of internal fluctuations cannot be *a priori* neglected [21]. As a matter of fact, as shown below, they can play a key role for SR in realistically small isolated clusters of ion channels like in Fig. 11.2.

11.3.1 Quantifying Channel Noise

The role of channel noise for the neuron firing has been first studied by Lecar and Nossal as early as in 1971 [28]. The corresponding stochastic generalizations of Hodgkin-Huxley model (within a kinetic model which corresponds to the above given description) has been put forward by DeFelice *et al.* [29] and others; see [21] for a review and further references therein. The main conclusion of these previous studies is that the channel noise can be functionally important for neuron dynamics. In particular, it has been demonstrated that channel noise alone can give rise to a spiking activity even in the absence of any stimulus [21, 29, 30], see also in Fig. 11.3.

To include the channel noise influence in a theoretical modeling within the stochastic kinetic schemes [21, 29], however, necessitates extensive numerical simulations [31]. To aim at a less cumbersome numerical scheme we use a short-cut procedure that starts from (11.4) in order to derive a corresponding set of Langevin equations for a stochastic generalization of the Hodgkin-Huxley equations of

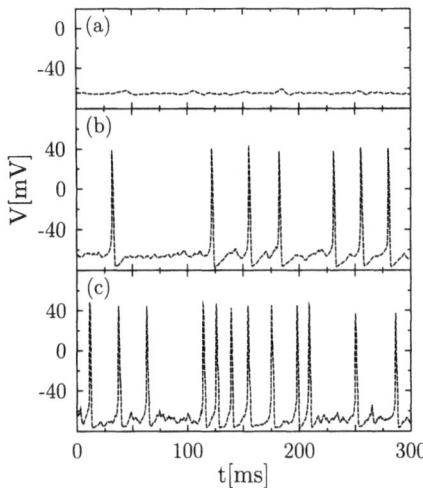

Fig. 11.3. Numerical simulation of the stochastic Hodgkin-Huxley system (11.1), (11.6), (11.7) with vanishing external stimulus. We computed several realizations of the voltage signal for different numbers of the ion channels: **a** $N_{Na} = 6000$, $N_K = 1800$; **b** $N_{Na} = 600$, $N_K = 180$; and **c** $N_{Na} = 60$, $N_K = 18$. Upon decreasing the system size the influence of channel noise on the spontaneous firing dynamics becomes more and more pronounced. Note that the non-stochastic Hodgkin-Huxley model does not exhibit spikes at all for the parameters given in the text and in the absence of external stimuli

the type put forward by Fox and Lu [32]. Following their reasoning we substitute the equations (11.4) with the corresponding Langevin generalization:

$$
\begin{aligned}
\dot{m} &= \alpha_m(V)\,(1-m) - \beta_m(V)\,m + \xi_m(t)\,, \\
\dot{h} &= \alpha_h(V)\,(1-h) - \beta_h(V)\,h + \xi_h(t)\,, \\
\dot{n} &= \alpha_n(V)\,(1-n) - \beta_n(V)\,n + \xi_n(t)\,,
\end{aligned}
\tag{11.6}
$$

with independent Gaussian white noise sources of vanishing mean. The noise autocorrelation functions depend on the stochastic voltage and the corresponding total number of ion channels as follows:

$$
\begin{aligned}
\langle \xi_m(t)\xi_m(t')\rangle &= \frac{2}{N_{Na}}\,\frac{\alpha_m\beta_m}{(\alpha_m+\beta_m)}\,\delta(t-t')\,, \\
\langle \xi_h(t)\xi_h(t')\rangle &= \frac{2}{N_{Na}}\,\frac{\alpha_h\beta_h}{(\alpha_h+\beta_h)}\,\delta(t-t')\,, \\
\langle \xi_n(t)\xi_n(t')\rangle &= \frac{2}{N_{K}}\,\frac{\alpha_n\beta_n}{(\alpha_n+\beta_n)}\,\delta(t-t')\,.
\end{aligned}
\tag{11.7}
$$

In order to confine the conductances between the physically allowed values between 0 (all channels are closed) and g^{max} (all channels are open) we have implemented numerically the constraint of reflecting boundaries so that $m(t)$, $h(t)$ and $n(t)$ are always located between zero and one [32].

Moreover, the numbers N_{Na} and N_K depend on the actual area S of the membrane patch. With the assumption of homogeneous ion channels densities, $\rho_{Na} = 60\ \mu\mathrm{m}^{-2}$ and $\rho_K = 18\ \mu\mathrm{m}^{-2}$, the following scaling behavior follows:

$$
N_{Na} = \rho_{Na}S\,, \quad N_K = \rho_K S\,.
\tag{11.8}
$$

Upon decreasing the system size S, the fluctuations and, hence, the internal noise increases. Consequently, with abating cell membrane patch the spiking behavior changes dramatically, cf. Fig. 11.3.

The numerical integration is carried out by the standard Euler algorithm with the step size $\Delta t \approx 2 \cdot 10^{-3}$ ms. The "Numerical Recipes" routine ran2 is used for the generation of independent random numbers [33] with the Box-Muller algorithm providing the Gaussian distributed random numbers. The total integration time is chosen to be a multiple of the driving period $T_\Omega = 2\pi/\Omega$, as to ensure that the spectral line of the driving signal is centered on a computed value of the power spectral densities. From the stochastic voltage signal $V(t)$ we extract a point process of spike occurrences $\{t_i\}$:

$$u(t) := \sum_{i=1}^{N} \delta(t - t_i), \qquad (11.9)$$

where N is the total number of spikes occurring during the elapsed time interval. The occurrence of a spike in the voltage signal $V(t)$ is detected by upward-crossing a certain detection threshold value V_0. Obviously, the threshold can be varied over a wide range with no effect on the resulting spike train dynamics.

The power spectral density of the spike train (PSD_u) has been analyzed in the absence and in the presence of periodic stimulus and noise. In order to quantify SR, we obtain from the PSD_u the spectral power of the transmitted periodic signal, η, as the difference between the peak value of the spectral line and its background offset located at the driving frequency Ω. The another important measure, signal-to-noise ratio (SNR), is then given by the ratio of the spectral power of signal to the background offset (in the units of spectral resolution of signals).

11.3.2 Stochastic Resonance

First, we focus our attention on SR *in absence of external noise*, see Fig. 11.4a and Fig. 11.5a. Here, we discover the novel effect of genuine *intrinsic stochastic resonance*, where the response of the system to the sub-threshold external stimulus is optimized *solely* due to internal, ubiquitous noise. For the given parameters, SR in the spectral amplification of signal occurs at $S \approx 10 \ \mu m^2$ and in the signal-to-noise ratio at a different value $S \approx 32 \ \mu m^2$. Starting from $S \approx 10$ μm^2, growing internal noise monotonically deteriorates the amplitude of system response at the signal frequency. Moreover, upon reaching $S \approx 32 \ \mu m^2$ it deteriorates the quality of signal transduction which is measured by SNR. In this respect, it is worth mentioning that SNR also measures in effect the rate of information transfer [34], but for small-amplitude signals only [35].

Under such circumstances, one would predict that the addition of an external noise (which corresponds to the conventional situation in biological SR studies) *cannot* improve η and SNR further, *i.e.* conventional SR will not be exhibited. In order to verify this prediction, we contaminated the periodic stimulus (11.5)

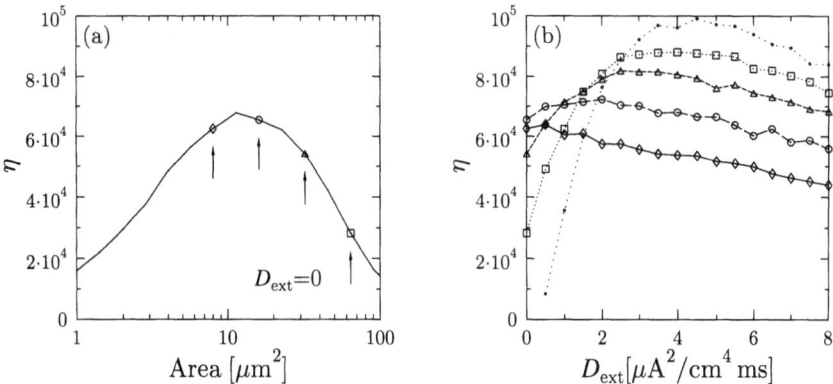

Fig. 11.4. The spectral amplification η of an external sinusoidal stimulus with amplitude $A = 1.0\ \mu\text{A/cm}^2$ and angular frequency $\Omega = 0.3\ \text{ms}^{-1}$ for different observation areas: **a** no external noise is applied; **b** SNR versus the external noise for the system sizes indicated by the *arrows* in Fig. 11.5a: $S = 8\ \mu\text{m}^2$, *solid line through the diamonds*; $S = 16\ \mu\text{m}^2$, *long dashed line connecting the circles*; $S = 32\ \mu\text{m}^2$, *short dashed line through the triangles*; $S = 64\ \mu\text{m}^2$, *dotted line connecting the squares*. The situation with no internal noise (i.e., formally $S \to \infty$) is depicted by the *dotted line connecting the filled dots*

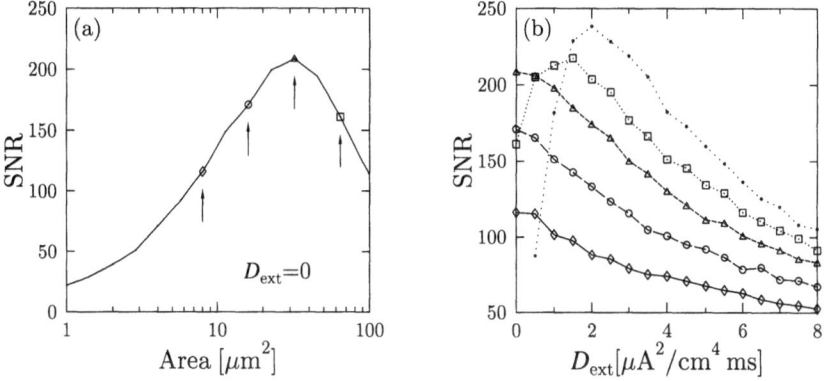

Fig. 11.5. The signal-to-noise ratio (SNR) for the same parameters as in Fig. 11.4

by the addition of Gaussian white noise $\zeta(t)$. The latter one possesses the autocorrelation function

$$\langle \zeta(t)\zeta(t')\rangle = 2D_{\text{ext}}\ \delta(t - t')\,, \tag{11.10}$$

and the noise strength D_{ext}. The corresponding results, depicted in Fig. 11.4b and Fig. 11.5b, fully confirm the above prediction. Conventional stochastic resonance therefore occurs only for large membrane patches beyond optimal sizes and reaches saturation in the limit $S \to \infty$ (limit of the deterministic Hodgkin-Huxley model). Thus, the observed biological SR [10, 11] is rooted in the collec-

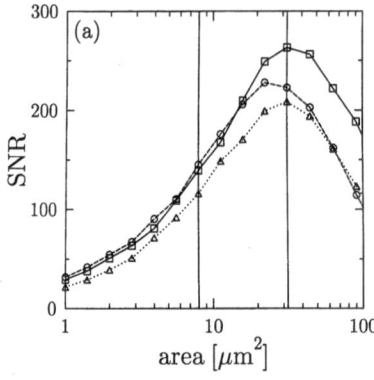 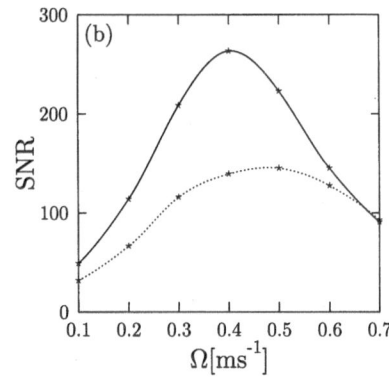

Fig. 11.6. The signal-to-noise ratio (SNR) for a sub-threshold external stimulus with amplitude $A = 1.0$ μA/cm^2 and different angular frequencies: **a** SNR versus the observation area for $\Omega = 0.3$ ms^{-1} (*dotted line through the triangles*), $\Omega = 0.4$ ms^{-1} (*solid line connecting the squares*), and $\Omega = 0.5$ ms^{-1} (*dashed line through the circles*); **b** SNR versus the driving frequency for two areas ($S = 8$ μm^2, *dotted line*; $S = 32$ μm^2, *solid line*), depicted by vertical lines in Fig. 11.6a. The curves exhibit clear maxima and, therefore, a combination of the stochastic resonance with the conventional frequency resonance takes place

tive properties of large ion channels arrays, where ion channels are globally coupled via the common membrane potential $V(t)$.

In addition, by changing the driving frequency we rediscover the effect of combined stochastic resonance and conventional resonance [20, 26], cf. Fig. 11.6. In other words, SNR becomes optimized not only versus the patch size, but also versus the driving frequency. Moreover, due to the noisy character of gating variables, the mean frequency of a corresponding random limit cycle in the stochastic Hodgkin-Huxley model (11.1),(11.6),(11.7) depends on the membrane patch area. Thus, the maxima of SNR are located for various system sizes at different driving frequencies. This effect may be used for frequency tuning in the biological signal transduction.

11.4 Conclusions

In conclusion, we have investigated the stochastic resonance in a noisy generalization of the Hodgkin-Huxley model. The spontaneous fluctuations of the membrane conductivity due to the individual ion channel dynamics has systematically been taken into account. We have shown that the excitable membrane patches exhibit a spontaneous spiking activity due to the omnipresent internal noise.

The main result of this study refers to the phenomenon of *intrinsic SR*. Here, the channel noise *alone* gives rise to SR behavior, cf. Fig. 11.4a and Fig. 11.5a (see also [31]). Moreover, such intrinsic SR becomes optimized versus the driving angular frequency, cf. Fig. 11.6. Conventional SR versus the external noise intensity also takes place, but for sufficiently large membrane patches, where the internal noise strength alone is not yet at its optimal value. We thus conclude that the observed biological SR likely is rooted in the *collective* properties of globally coupled ion channel assemblies.

The authors gratefully acknowledge support for this work by the Deutsche Forschungsgemeischaft, SFB 486 *Manipulation of matter on the nanoscale*, project A10.

References

1. D.J. Watts and S.H. Strogatz, Nature (London) **393**, 440–442 (1998)
2. A.L. Barabasi and R. Albert, Science, **286**, 509–512 (1999)
3. R. Albert and A.L. Barabasi, Rev. Mod. Phys. **74**, 47–98 (2002)
4. S.N. Dorogovtsev and J.F.F. Mendes, Adv. Phys. **51**, 1079–1187 (2002)
5. R. Cohen, K. Erez, D. ben-Avraham and S. Havlin, Rev. Phys. Lett. **85**, 4626–4628 (2000)
6. R. Cohen, K. Erez, D. ben-Avraham and S. Havlin, Rev. Phys. Lett. **86**, 3682–3685 (2001)
7. R. Cohen, D. ben-Avraham and S. Havlin, Rev. Rev. E. **66**, 036113 (2002)
8. L. Gammaitoni, P. Hänggi, P. Jung, and F. Marchesoni, Rev. Mod. Phys. **70**, 223–288 (1998)
9. V.S. Anishchenko, A.B. Neiman, F. Moss and L. Schimansky-Geier, Usp.Fiz. Nauk **169**, 7–38 (1999) [Physics – Uspekhi **42**, 7–36 (1999)]
10. J.K. Douglass, L. Wilkens, E. Pantazelou, and F. Moss, Nature (London) **365**, 337–340 (1993)
11. J.E. Levin, and J.P. Miller, Nature (London) **380**, 165–168 (1996)
12. J.J. Collins, T.T. Imhoff, and P. Grigg, Nature (London) **383**, 770 (1996)
13. E. Simonotto, M. Riani, C. Seife, M. Roberts, J. Twitty, and F. Moss, Phys. Rev. Lett. **78**, 1186–1189 (1997)
14. I. Hidaka, D. Nozaki, and Y. Yamamoto, Phys. Rev. Lett. **85**, 3740–3743 (2000)
15. S.M. Bezrukov, and I. Vodyanoy, Nature (London) **378**, 362–364 (1995); **385**, 319–321 (1997)
16. I. Goychuk, and P. Hänggi, Phys. Rev. E **61**, 4272–4280 (2000)
17. D. Petracchi, M. Pellegrini, M. Pellegrino, M. Barbi, and F. Moss, Biophys. J. **66**, 1844–1852 (1994)
18. A.L. Hodgkin, and A.F. Huxley, J. Physiol. (London) **117**, 500–544 (1952)
19. A. Longtin, J. Stat. Phys. **70**, 309–327(1993);
    K. Wiesenfeld, D. Pierson, E. Pantazelou, C. Dames, and F. Moss, Phys. Rev. Lett. **72**, 2125–2129 (1994);
    J.J. Collins, C.C. Chow, A.C. Capela, and T.T. Imhoff, Phys. Rev. E **54**, 5575–5584 (1996);
    S.-G. Lee, and S. Kim, Phys. Rev. E **60**, 826–830 (1999)
20. B. Lindner, and L. Schimansky-Geier, Phys. Rev. E **61**, 6103–6110 (2000)
21. J.A. White, J.T. Rubinstein, and A.R. Kay, Trends Neurosci. **23**, 131–137 (2000)

22. B. Hille, *Ionic Channels of Excitable Membranes*, 2nd ed. (Sinauer Associates, Sunderland, MA 1992)
23. B. Sakmann, and E. Neher, *Single-Channel recording* (Plenum Press, 1995)
24. R.J. Nossal, and H. Lecar, *Molecular and Cell Biophysics* (Addison-Wesley, Redwood City 1991)
25. S.B. Lowen, L.S. Liebovitch, and J.A. White, Phys. Rev. E **59**, 5970–5980 (1999)
26. Y. Yu, W. Wang, J. Wang, and F. Liu, Phys. Rev. E **63**, 021907 (2001)
27. N.P. Issa and A.J. Hudspeth, Proc. Natl. Acad. Sci. USA **91**, 7578–7582 (2001)
28. H. Lecar, and R. Nossal, Biophys. J. **11**, 1068–1084 (1971)
29. J.R. Clay, and L.J. DeFelice, Biophys. J. **42**, 151–157 (1983);
    A.F. Strassberg, and L.J. DeFelice, Neural Comput. **5**, 843–855 (1993);
    L.J. DeFelice, and A. Isaac: Chaotic states in a random world, J. Stat. Phys. **70**, 339–354 (1993)
30. G. Schmid, and I. Goychuk, P. Hänggi, Europhys. Lett. **56**, 22–28 (2001)
31. P. Jung, and J.W. Shuai, Europhys. Lett. **56**, 29–35 (2001)
32. R.F. Fox R, and Y. Lu, Phys. Rev. E **49**, 3421–3431 (1994)
33. W.H. Press, S.A. Teukolsky, W.T. Vetterling, and B.P. Flannery, *Numerical Recipes in C*, 2nd ed. (Cambridge Univ. Press, Cambridge 1992)
34. F. Rieke, D. Warland, R. de Ruyter van Steveninck, and W. Bialek, *Spikes: Exploring the Neural Code* (MIT Press, Cambridge, MA, 1997)
35. I. Goychuk, Phys. Rev. E **64**, 021909 (2001)

Lecture Notes in Physics

For information about Vols. 1–582
please contact your bookseller or Springer-Verlag
LNP Online archive: http://www.springerlink.com/series/lnp/

Druck: betz-druck GmbH, D-64291 Darmstadt
Verarbeitung: Buchbinderei Schäffer, D-67269 Grünstadt